CW01523370

While writing the final pages of this book, I lost two of my dearest friends;
I dedicate these pages to them,
For Juliet for allowing me to take refuge in your kitchen twenty five years ago.
For Ross for sharing your knowledge and giving me the confidence to make wine.
"Wishing to be friends is quick work, but friendship is a slow ripening fruit"
Aristotle (4th century BC)

I would like to thank Ceri Prenter of Sunbird Publishers for making
this book possible, Russel Wasserfall for his hours spent editing, his valuable
creative input and magnificent photography, Roxy for the beautiful design
and my family for their undying support and unfailing appetites.

SUNBIRD
PUBLISHERS

Harvest diaries

Christine Stevens

Foreword

Having spent twenty years touring the globe as a fashion buyer, I finally settled with my husband and two young sons on a beautiful old farm nestled in the mountains of the Western Cape. This old farm, long neglected before we arrived, has woven itself completely into the fabric of my life.

When we started looking for a piece of land, it was always our intention to farm organically, and so it followed that in establishing the winery, we worked at using time-honoured, natural methods. Wine was not the only thing that was allowed to develop naturally. Drawn into the rhythm of life dictated by the seasons, my kitchen table soon heaved under the weight of piles of our home-grown organic produce. As I became more absorbed in growing and producing healthy delicious meals for my family, the vegetable and herb gardens began to spread outwards from the kitchen door.

My love affair with food must have started as I played in the fruit orchard of my parents' garden as a young girl, then my senses were assaulted as I ate my way through Italian and French bistros after a hard day's work. Travelling days behind me, my food journey reached its climax here on this old Cape farm, where I have given in to the sheer joy of picking my own vegetables and fruit, bringing them into the kitchen, and transforming them into delicious meals. My happiest evenings are without doubt spent enjoying the rewards of our harvest, with family and friends.

First published in 2010

Sunbird Publishers (Pty) Ltd

(A division of Jonathan Ball Publishers (Pty) Ltd)

PO Box 6836, Roggebaai 8012

Cape Town, South Africa

www.sunbirdpublishers.co.za

Registration number: 1984/003543/07

Copyright published edition © Sunbird Publishers

Copyright text © Christine Stevens

Copyright photographs © Russel Wasserfall

Design & layout by Roxanne Spears

Edited by Margaret Wasserfall

Proofread by Sarah Kate Schafer and Kathleen Sutton

Publisher Ceri Prenter

Reproduction by Resolution Colour (Pty) Ltd, Cape Town

Printed by Tien Wah Press (Pte), Singapore

ISBN 978-1920289-26-3

Contents

January

Every year, I move my office outside for the month of January, and there it stays. It's really quite a simple affair. I am content with a simple green metal table under the tall pin oak that towers over the swimming pool. This tree is not as old as the huge European oaks growing close by, but I love the delicate green leaves that rustle gently in the summer breeze. It's the perfect location, as the pool area becomes the hub of summer living for the family. It's a place of happy memories and a few bumpy landings.

When we first moved to the farm ten years ago, this tree was the willing shade house for the delicate cuttings that needed protection before I could establish a permanent shade nursery. It is also the tree that holds painful memories for my eldest son. After a couple of summers frolicking in the pool he fixed his eye on the tree and decided that a slide plunging from its wide branches into the pool below would be just the ticket. The tree was rather unco-operative and he came tumbling down, denting body and dignity.

Each morning I carry my laptop, papers and coffee outside to my makeshift desk. There are few offices with such a stunning view and it is blissful working under the canopy of leaves where a cool breeze is nature's air conditioner. The Slanghoek Valley snakes between two mountain ranges which can be temperamental and unpredictable. It's not unknown for summer storms to strike this valley in an instant.

January is full of surprises. In the first year after we bought the farm, we were celebrating New Year's Eve when a storm blew through this seemingly tranquil place bringing hail that devastated the ripening grapes. A couple of years later a strong north-westerly wind blew over row upon row of grape-laden vines.

My vista is superb. Rows of vines are backed by imposing mountains. Bunches of jasper-green grapes hang from the vines. I am waiting for the first blush of pale pink that will herald the rich purple of fully ripe grapes. This year, the process has been slower and I know this is due to storms that raged through the valley in November. I will have to be patient.

Peace reigns in the mountain-rimmed valley. The farm workers are on their annual summer vacation, before the hectic routines of grape harvest begin. Opposite me, the horses graze contentedly in their paddocks. I can work uninterrupted. Apart from the comforting sing-song of the birds and the gentle sounds of our animals, the farm is silent. I write, research and answer correspondence in the cool of the morning, before the heat of the day brings children to play in the pool, and demands for food are issued.

This morning I took an early walk down to the vineyard to inspect the vines for that telltale blush of ripening pink. There is not a breath of wind and the day will be hot. I turn on water for the thirsty vegetable garden, and then stroll over to say good morning to our hens. They dislike this heat and I can tell from their dropping feathers and the clucking banter between them that they are beginning to feel the discomfort of the day. They are going to need more drinking water and I turn up the supply tap. Egg production will be affected if this carries on.

Twenty-one chicks were hatched this year. One poor mother sat successfully on fourteen eggs. What patience and dedication. The chicks are now twelve weeks old and getting stronger, running around with such speed that I wonder how that mother manages to keep an eye on them all at once. Her biggest problems are the pied crows that wheel in the skies by day, and the sly genets that raid the nests by night. I spare a thought for the millions of less fortunate battery hens, who never see daylight or get to scratch the earth for tasty treats.

The peach tree next to the nursery paddock where the hens live is laden. Several peaches have been pecked at by the birds, which always seem to go after the ripe fruit first. This morning I beat them to it and picked eight glorious fresh peaches for breakfast, devouring two on my way to the kitchen.

Cooking is such a pleasure at this time of year. The vegetable gardens are abundant with offerings. Simple meals made from fresh ingredients are eaten outside. Breakfasts are served mid-morning when everyone has surfaced. Supper is eaten later in the evening when we take advantage of the long light evenings to sit outside under a beautiful starlight sky. This is the best way to share food and wine, as the taste just seems to get better when food is eaten outdoors. The pace is leisurely and there are no pressures of early morning commitments.

This is the season when fruit is served with every meal, in one form or another. Berries ripen and peaches, plums, figs and cherries hang from the trees, like edible Christmas decorations. Huge juicy melons trail along the ground, growing larger and riper under the sun. These sweet ripe globes naturally detach themselves from the stem when ready to be eaten. In summer, sweet melons are included in nearly every meal. I serve them on their own, add them to salads, garnish them sliced with slivers of air-dried hams and salami. My sons love them cooked with duck, glazed by their honeyed juices at the end of cooking.

Yesterday I gathered a basket of juicy ripe specimens and left some outside the kitchen. This morning I saw that bites had been taken from three of them. I'm sure it must have been a baboon. I often hear them barking on the mountainside at night. Last year they raided through rows of my ripe sweetcorn, taking bites from several cobs and discarding them. I always used to wonder why they

never finish one until an ecologist explained that this is nature's way of feeding everyone. The fruit dropped by baboons is then eaten by the creatures that are unable to reach it themselves. Duikers, small buck, bush pigs, porcupines and insects all benefit from the baboon's leftovers. I display the damaged fruit at breakfast. Mark thinks it might be baboons; one of the boys thinks it could be meerkats. Both creatures are diurnal and meerkats are insectivorous, so that seems to rule out both of these culprits. Could it have been a porcupine? I cannot understand why we were not alerted to this intruder by the dogs. We all agree to keep a careful watch over the fruit and vegetable gardens until we identify our visitor.

Our strawberry beds started off as fifteen small runners brought to me by a neighbour many years ago. Ever since then, every two years, I detach new runners and replant them. From then on they pretty much look after themselves. Our beds have now grown out to several hundred trouble-free plants producing new fruit every day, which I can pick on most evenings. We eat the best fruit for desserts and every few days the remaining berries are made into jam to store and enjoy during the cold months to follow.

The strawberry beds are planted between the rows of apple trees in the orchard. When we moved to the farm, I was told that apples would never grow in this area. I was stunned by this piece of information, as our winters are certainly cold enough for apples. So, I planted three trees anyway and they seemed to flourish. Two years later we planted fifty more to form an orchard. We toiled one cold winter's day measuring out rows, planting and staking young trees with old vine poles. There's a strong north-westerly that howls through our valley at such an alarming speed it will unearth anything in its path that is not firmly staked.

The treasured apple trees reward us year after year with crisp red fruit, so sweet that one year our stallion broke out of his paddock in the dead of night to feast on the sweet apples. It didn't take long to work out who was responsible for this loss. The telltale evidence of huge hoofprints could not be hidden.

I have planted rows of tomatoes next to the strawberry beds. Over the years I have collected and saved the seeds of many different varieties. What would we do without this wonderful fruit, which, as legend tells us, was discovered by Cortez, the Spanish explorer? He stumbled upon the plants in Montezuma's garden. From that legendary moment the tomato has been a flavoursome addition to the world's culinary traditions. It is clearly a firm favourite in our household.

At this time of year, the tomato, like the melon, will accompany every meal. Small round ones are chopped up with sweet red onions for a salsa. Huge beefsteak tomatoes, the size of my hand, are finely sliced and served with generous handfuls of fresh herbs and lashings of olive oil.

Rows of garlic grow alongside the tomatoes. These are the descendants of a huge globe that I brought back from France one year. It was so much bigger than the local garlic that I simply could not resist it. After enjoying several cloves, I tossed a couple into the soil. The new garlic shot up and I was thrilled. Every year, when harvesting, I save some cloves to replant. Growing garlic is not for impatient souls. It takes nine full months, the amount of time it takes for a human baby to form, from sowing to being able to harvest your crop.

The evening is beautifully still. This is a rare thing here where the wind usually whips through the valley each evening. It will wait for you to set the table for dinner before disrupting proceedings. Warm roast chickens are on the supper menu. They've been stuffed with herbs, to be served just at room temperature. They are moist and juicy with crisp skin crusted with fresh herbs and salt. The chicken is accompanied by a tart of tomatoes, baked on puff pastry, brushed with French mustard and seasoned with garlic and oregano. This is simplicity itself, as the tart can be made with shop-bought or homemade puff pastry. It is important to let the tart rest and cool a little before serving, because the tomatoes will be quite runny when you take the tart from the oven and hot tomatoes taste of nothing. They will firm up when rested. A simple supper eaten with gusto. Plates are mopped clean with the fresh bread. Who's going to clear the dishes and load the dishwasher? We all scuttle around then settle down to fresh strawberries and some local cheese.

It's almost a shame to waste a glorious night by sleeping. But, after a quick dip in the pool I flop into bed. The house is quiet. Everyone is asleep. I hear the hoot of an owl as my eyes close on a beautiful summer's day.

In January the wine lies still in its barrels, and I regularly check on it all. The barrels must be topped up as small amounts of wine will evaporate. Air is the enemy of wine, making it susceptible to oxidation and spoilage. The cellar is cool, chilled to keep a stable temperature, so much so that when you walk in from the outside the coolness hits you like a cold slap.

My heart sinks as the holidays draw to an end. My sons are weekly boarders, returning home for weekends. I miss them. They are such an important part of the farm. Like the weft on which a tapestry is woven, they are intrinsic to its formation and growth. Both are so actively involved in the farm that when they are away it feels incomplete, like someone has taken a huge bite out of life.

Slowly I adapt to cooking for two. For the first couple of days miscalculations occur in my food planning and the fridge shelves fill up with leftovers. My new routine slips into place and, with new energy levels, the lazy days are left behind as the stride of the new year picks up its pace.

Our staff return from annual leave and farm activities commence. I disappear into the almond orchard with our five ladies armed with the long sticks they use to beat the almonds from the trees. These almond trees are very old. Their girths are irregular and they lean to one side like tired old men. One winter a splendid old tree was blown over, but, with the help of two men and a tractor we pulled that tree up and supported it with wire stays covered in rubber piping. I hoped it might survive, but it never recovered. The trunk was loaded onto a trailer, and it now lies outside my office, a sculpture of nature. Smaller branches were cut up for firewood. The wood was as hard as iron and gave off a tremendous heat on the fire.

Almonds originated in Asia, and then spread to the Mediterranean. How they arrived in Africa is a mystery I have yet to solve, but I am eternally grateful to whoever planted these beautiful specimens on our farm all those years ago. The nuts are ready to harvest when their furry outer shells start to dry and split open revealing the treasured nuts inside. Once the ripe nuts have been brought down to earth we gather them up into harvest buckets. We're careful to leave those that are not totally ripe for the next gathering. Almond orchards must be planted with two different varieties to ensure cross-pollination and therefore a crop. One variety ripens first and we leave the rest to fully ripen on the trees.

Once gathered, the outer covering must be pealed from the nut, which has to be dried. We place our almonds on the flat feed store roof, which gets pretty hot. So it's not long before these almonds are on our menu, tossed in a little olive oil and salt, cooked in a shallow-based pan. I also grind blanched almonds in a blender to use as a substitute for flour in cakes and tarts.

It is the end of January, and temperatures start to climb. Even the dogs choose to lie on the cool tile floor inside. Wine buyers arrive from Europe to taste our wines. They have come from the south of England where temperatures are -15 degrees Celsius. Here it's topping 38 degrees. Is this the effect of global warming? I wonder how this will affect our farm.

Even with this heat, the crops are late. The cold, wet spell of November has upset the rhythm of the seasons. The vines are enduring the heat of the day. We do not irrigate our vines, believing it better for them to develop a deep root system that sips at the cool moisture to be found deep underground. I know too that they will be refreshed by the cool of the night.

Early morning dew drops hang from the leaves like crystal spangles. As the morning progresses the temperature rises until the cool of the dawn is a distant memory. Each day the grapes ripen more. The chardonnay start to yellow, the red grapes grow a deeper shade of purple.

Each month brings different tastes to my tongue. January's flavour is tomato, fresh basil, mozzarella, and a splash of olive oil. To satisfy my taste buds I pick some large round beefsteak tomatoes to make this salad for lunch. I always interplant basil with the tomatoes and now I pluck a few leaves as I walk down the rows. Plants that taste good together tend to grow well together.

A mid-morning walk in the vineyards is foolhardy if you don't have a hat. The heat intensifies each day, drying out the earth and sapping the fresh green from the oak leaves. A film of fine dust rises like a dry wake behind vehicles on the dirt roads. My beloved roses start to fade in the heat; they will have a second flush close to Easter, when the weather cools. The old English varieties have faded and will not bloom for another year. I pick the last few flowers for a vase in my bedroom. When I go to bed later that night their scent fills the room.

The vegetable gardens demand more water as lettuce and spinach wilt, giving up the fight to produce leaves. It's impossible to grow tender crops now, but tomatoes, courgettes and peppers are completely happy. In the evenings I water pots, which dry out easily in the heat. I resort to moving the potted basil to a shady spot when I notice that it is starting to cover itself in the small white flowers that will ruin the flavour of my pesto. I pluck them off quickly to better the odds.

In the heat of the afternoon the horses love to be sprayed down with a hose. They arch their backs, stretch their necks and toss their heads with enjoyment. Another cooling treat for them is to be taken down to the river, where they frolic in the shade of the weeping willows. This is a horsey highlight for the day and whoever leads them down to the river is guaranteed a soaking.

I love my late evening summer walks, when the air is cooler and the sky is invariably illuminated by a trillion stars. My route is planned to pass the tap so I can switch off the sprinklers in the vegetable patch. They are watered in the late afternoon so that the moisture is absorbed by plants and soil and not evaporated in the heat of the day, damaging delicate leaves. The walks let me size up what's growing or ripening, but they also ensure that I always sleep well at this time of the year. If we have family or friends staying in our guest cottage, they often walk with me, stopping to gape and breathe in the cool, pure night air.

February approaches and summer is once again in full swing.

Tomato Tart

500g	tomatoes, small or cherry
1 tbsp	olive oil
handful	oregano leaves, roughly torn
	sea salt to taste
1	sheet of butter puff pastry
1 tbsp	French Dijon mustard

Serves 6

Preheat the oven to 180°C or gas mark 4. Cut the tomatoes in half and place them in a shallow pan with the olive oil. Simmer on a low heat for 5 minutes, until the tomatoes have just started to soften. Toss in the oregano and a pinch of salt, and leave to rest on one side. Take the sheet of puff pastry and place on a sheet of baking paper onto a baking tray. Spread the mustard over the pastry, then spoon the tomato mixture on top, spread the tomatoes out thin to cover the pastry, leaving about a centimetre around the edges. Place the tray in the oven and bake for 20 minutes, or until the edges of the pastry are crispy and brown. Let the tart cool for about 10–15 minutes. Cut into squares and serve.

Salted Herbed Almonds

Toss blanched almonds with a little olive oil over a low heat in a frying pan until pale brown. Don't walk away as they burn easily. Pour onto a plate covered with a piece of kitchen towel to absorb the excess oil. While the almonds are still hot, flavour with sea salt and a pinch of dried herbs. Serve cold with a glass of wine. I also toss them into a green salad sometimes. Any leftovers will keep in a jar for 2 to 3 weeks.

Duck with Melon

6	duck breasts
125ml	white wine
1 tbsp	honey
	salt and pepper to taste
6	thin slices of sweet melon

Serves 6   Place a shallow large frying pan on the stove and heat it up for about 3 minutes.
When the pan is hot, place the duck breasts in the pan skin side down. They will make
a sizzling noise. Leave for 3–5 minutes until the duck skin has turned golden brown.
At this point it's a good idea to pour off the fat that has come out of the skin.
When the skin is golden brown, turn the duck breast over and reduce the heat slightly.
Cook for a further 5 minutes, a little more if the breasts are thick or if you like them
well done. Remove the breasts from the pan and put to one side to rest on a dish.
Add the white wine to the pan and put the heat back up, then add the honey and let
the juice reduce for a few minutes. Add a pinch of salt and pepper to season; you can also
add the juice that comes out of the resting duck breasts. The juices will thicken slightly,
add to this the melon slices, then turn off the heat, so they become glazed but do not cook
and go soft. Turn the melon over in the juices to coat it, but do not cook for long as it
will go soggy. The melon should still be firm. Slice each duck breast at an angle into three.
Place the breasts onto a plate with a slice of melon and spoon over the juice.
Serve with a green salad and fresh crusty bread.

Almond & Plum Tart

For the rich shortcrust pastry

250g	plain flour
175g	unsalted butter
1	large egg, beaten

For the frangipane filling

300g	unsalted butter, softened
300g	organic unrefined white sugar
300g	ground almonds
3	large eggs, lightly beaten
6	ripe plums

Serves 8-10

Preheat the oven to 180°C or gas mark 4. For the pastry, work the flour into the butter until the mixture resembles crumbs. Add the beaten egg to bind the pastry. If it is a little dry, add a little cold water. Roll out the pastry and place into a lightly buttered 28cm metal tart tin with a loose base. Let the pastry rest for 10 minutes then blind bake for 10 minutes. To make the frangipane, cream together the butter and sugar, add the almonds and lastly the eggs. Once the tart shell has been blind baked, spread all the frangipane into the tart. Cut your six plums in half and arrange them cut side down on the filling. They will sink in slightly.

Place the tart back in the oven and cook for a further 25–30 minutes or until the tart is a pale golden brown. Cool slightly before serving. Serve with cream... It is also delicious cold.

Tip: You can substitute the plums with apricots, or any of the soft summer fruits such as cherries, berries or even pears. This tart can be eaten warm or cold and will keep for three to four days in the fridge.

February

The wind howls all night, keeping me awake. I rise early and go down to the kitchen to make a pot of coffee. While I wait for the telltale bubbling of the coffee pot, I look out over the mountains in front of the house. Grey clouds block the top of the mountains from view. The heat has broken. Cooler weather is on its way, for a few days at least. It will be full moon soon, and this often brings a change of weather. In the winter months I've noticed that our worst storms always fall around the time of the full moon.

Each day I inspect the grapes. Harvest is drawing closer. The chardonnay grapes are becoming opaque as they ripen, yellowing slightly. The reds are already turning to the deep purple that will signify their readiness.

This is a critical time and we must be vigilant, removing any green grapes from the red bunches. These will never ripen correctly and, if they get into the press, they'll make the wine tart and acidic. It's time-consuming work. Every underdeveloped green grape is gently prised off the bunch. It takes ages, but it's vital to the smooth taste of our wines.

During the run-up to harvest it's also essential to regularly check the sugar and acid readings of the grapes. If you can pick your grapes at exactly the right moment, your wine will taste sublime. South African winemakers are allowed to add tartaric acid to wine because it's difficult to maintain a healthy acid in the grape if the weather is too hot, but I hate this practice and avoid adding acid if possible.

Adding stuff like acid to wine is not natural and I think that it does affect the flavour of a good wine. Winemakers in France are permitted to add sugar, as they have the opposite problem to us – not enough sun.

As the days roll on, I start to walk through the chardonnay vineyard more and more frequently, randomly picking grapes from various bunches. Today I pop a little squirt of juice onto the lens of my small hand-held refractometer, which measures sugar content, and this confirms my suspicions. The chardonnay has a while to go yet.

Since the beginning of January we have had a steady stream of clients and friends visiting from Europe. It's lovely to see them and to entertain them but now I have only a few days left to catch up on those jobs that have been pushed to one side. There is no internet signal. The wind must have caused havoc somewhere. The admin must wait and I choose instead to work outside, in the vegetable gardens, tying up large tomato-laden vines. I lift the last remaining sweet melons from the ground and lay them on broken tiles to protect them from insects that burrow in where the fruit's skin has softened due to lying on the damp soil.

When we first moved to the farm I was over-joyed to find melons growing prolifically between the vines. Like an eager ingénue I picked one and rushed inside with my prize. Of course I was a laughing stock for the family that night at dinner, because at the first bite, our mouths were filled with the rank bitterness of wild melon. It was only much later, chatting to someone at the local co-op, that I discovered that the previous owner of the farm used to grow them for his cattle. I'm sure the word probably got around about my faux pas, and the bitter taste of that awful melon is one that is hard to forget.

It took years to eradicate the wild melons from the vineyards. Every year we'd collect every single one we could find until we thought there couldn't be any left. The next season another crop would appear, as if by magic. A melon contains so many seeds that it is easy for them to multiply. The happy part of the story though is that the bitter melon showed me just how easy it is to grow the delicious sweet variety. Just dry the seeds on paper towels laid out in the sun, then save them for the following year. That's the thing about growing your own food, nature provides a means for you to keep doing it. With all the stuff that goes on around GM seeds, I fear people might be stripped of their right to provide themselves with food from nature's pantry in a genetically modified world.

The wind has died down and the farm is still. All I can hear is the distant sound of the digger loader that my husband is driving somewhere at the far end of the farm. In the background is the constant chirruping of the crickets. Since it's cooler, I grab a fork and dig up my beloved French garlic, *allium sativum,* part of the same family as onions and leeks. Garlic is another species alien to our shores. Originating in Yunnan province in China, it has travelled far and wide across the world. The long green tubular leaves of my plants have turned brown and lie flat on the soil; a sign that the bulb is ready for harvest.

When they are fresh, the large bulbs tend to be streaked with deep pink and they smell heavenly. I leave them in the wheelbarrow to dry out over the next couple of days, then plait them into large strings, hoping this lot will last me the whole year. I take a few cloves to the kitchen to use straight away. Some of them will be broken up and planted in May for next year's crop.

Each bulb is made up of between twelve and fifteen cloves, so there should be plenty. I particularly love the taste of freshly dug young garlic and I use it to make a garlic mayonnaise, which is great mixed into a salad of new potatoes. I wonder why people use so many chemical and artificial flavourings in food when you can grow beneficial and superior natural flavours with such ease and minimal expense.

Next I attack the onions. The crop is not brilliant this year due to the heavy late rains that drowned them a bit. Like the garlic, the leaves of the onions have started to turn brown, and they must be dug out of the earth.

All this digging is not good for my back. The onions are dried in the sun, and in a few days I'll brush off the excess dirt and tie them into large bundles. There is a small room next to the house that we renovated to use as a food and vegetable store. The onions will keep well in there. I take a few of the freshly dug onions to the kitchen. My reward for all this digging will be a lovely, creamy onion tart for supper.

There is work to be done in the cellar. Last year's pinotage wine must be racked from its barrels. "Racking" is the term given to removing the lees and sediment that settles at the bottom of barrels during the year that it's been in the cellar. It's heavy and repetitive work, but it is part of the process of creating a drinkable wine. Basically what happens is that the wine comes out of the barrels and into a holding tank. Once the wine is in the tank and the barrels are empty, they are washed and rinsed. The wine is then pumped back into the barrels for the next stretch of maturation. This process involves lots of heavy lengths of pipe and loads of water.

A tank must be cleaned and ready to receive the wine, and water must be run through the pipes that are to be used. These pipes are about 15m long and are attached to a pump. When full of water or wine they are incredibly heavy. Carefully I insert a long, purpose-made nozzle attached to the pipe, or line as it is known in the industry, into the barrel and pump out the wine, watching to ensure that I don't pump out any of the sediments that lie at the bottom. The wine flows into the tank I've just prepared. After roughly twenty barrels are empty, they must be carried outside. Each one holds 225 litres so, even empty, they weigh an absolute ton. I have to enlist help for this part of the exercise as moving them is back-breaking work.

All the barrels are then thoroughly cleaned with copious amounts of water, and rinsed several times. When this is done we roll them back into position, securing them on their racks. Now we begin to refill them with wine from the holding tank.

Refilling must be done carefully to avoid spillage, and we have perfected a technique over the years. I just don't fill them too full from the pipe, and then I top up with a good old-fashioned jug. Job done, the tanks, the lines and the floor must be cleaned again.

By this stage, my arms usually feel as though they have been pulled out of their sockets. My hands look disgustingly stained and I am soaking wet. I cut a lemon and rub it all over my hands to remove the wine stain before sinking into a warm bath. On nights like these I will sleep like a baby.

The farmhouse is ringing with the voices of my sons, Nic and Sam, and a gaggle of their friends. A weekend whirlwind of activity is planned. They will fish and swim in the river, and spend hours in the shed fixing stuff.

Piles of dirty, wet clothes are bundled into the washing machine then hung out to dry on lines strung between the lemon trees. For my part, I'm happy as a lark cooking for hungry people. The batch of lemon muffins that I baked this morning have all been devoured and there's a rattle of tins and containers as the boys methodically clean out cookie jars and cake tins. Refilling the tins is a constant labour of love, especially during holidays or on summer weekends, when I know there will be a gang of hungry, growing boys foraging in the pantry. I've had to arm myself with a few quick and easy baking recipes to keep abreast of them.

Lamb chops from one of our own lambs are on the menu tonight. I plan to rub them generously with olive oil, shower them with herbs and jab slivers of garlic into them before cooking them quickly over the coals of a fire. Lamb with new potatoes and freshly picked green beans should fill them up for a while. After all the fervent activity of the day, everyone is tired and the house is dark and silent by ten o'clock.

We have only a handful of sheep, and a ram called Rambo. This chap has plenty of attitude. We take the lambs to slaughter between the age of 6 and 8 months. This has proven to be some of the tastiest and tender meat I have ever prepared. The lambs graze in the almond orchard where they enjoy a happy, stress-free life. However, when they go off to the butcher, I always feel sad. The men in my family are confirmed carnivores, but I have been troubled by the meat issue for thirty years. I feel happier eating our own meat, knowing its history.

A still Sunday morning is spent wandering around the garden picking up swimming towels and random articles of clothing abandoned by the boys around the pool. There is a cow happily helping herself to my vegetable garden, and I run towards her shouting like a demon. She looks at me in bovine horror and turns to flee back to the safety of the herd that is grazing contentedly in the river paddock.

Something, other than the cow, is eating my French beans. Whatever it is has feasted on the leaves, thankfully leaving the beans for us. But naked beans are a sorry sight. The baby carrots are intact so I'm guessing that the mystery raider is not partial to carrots. I make a note to myself to check if there's anything I can plant with the beans to deter things that go munch in the night.

As I head back into the kitchen, the smell of brioche dough reminds me to switch on the oven as there's baking to be done. This rich, cakey bread is an absolute indulgence for breakfast and is another recipe that I brought back from France. I always prepare the dough the night before as it needs to rise slowly at a low temperature. This one was not quite ready before I escaped into the garden this morning. The smell of freshly baked brioche must be one of the most comforting aromas on the planet, and it has the boys and their entourage of friends appearing from nowhere in search of sustenance. We breakfast on scrambled eggs, warm brioche, strawberry jam and coffee. Brioche tends to go stale very quickly, but if there's ever any left, it's also great toasted.

At this time of year, I constantly scan the weather reports and fret about the continuing heat. The chardonnay grapes are ripening fast, their sugar levels rise and their acids drop. It is slightly cooler, but the temperature is still over thirty degrees on most days. Threatening clouds gather overhead to be dispersed by the breeze that blows through the valley on most evenings.

The moment of peak ripeness arrives and we get up with the larks to harvest our first grapes of the season. Starting at first light, we pick while the air is still cool. Everyone concentrates on filling their lug boxes and soon the rows are stripped of their fruit, leaving the vines bare. I start the press and the juice runs into a tank – fresh cool grape juice that tastes so good.

The juice must settle in a cold tank while the big clean-up begins. The cement floor of the cellar is sticky from the stalks that have fallen; they are loaded onto a trailer heading for the compost heap. Everything is sprayed down well with water.

As new loads of grapes arrive at the cellar door, juice from previous days must be racked to clean tanks. Every day is filled with the endless tugging and cleaning of pipes and tanks. My arms ache, and I slosh around in wet boots. I'm the picture of glamour in grape-stained clothes.

Relieved that the first of the grapes are safely in the cellar, we sit on the front veranda sipping a glass of cold wine as we watch the sun slowly dissolve behind the mountains. Like a search light, the final ray of sunlight flashes through the deep gorge that splits the mountains in two. This is the cue for an evening performance of sound. First in the chorus line are the guinea fowl tink-tinking their strident note. Next a flock of Egyptian geese swoop overhead, disturbing everyone's song with their loud honking. The guinea fowl flap ponderously up into a tall oak tree and settle down for the night, squabbling amongst themselves for the best perch. All is quiet as darkness descends. I look up at a dazzling array of stars piercing the night sky and stroll barefoot up to the cellar to check the temperature of the juice settling in the tank. Tomorrow, this clear juice will be racked, or pumped, into another tank; small bubbles will appear as it starts to ferment and this clear nectar turns to wine.

One year we had a power failure during the night, which caused the cooling system in the cellar to shut down. Of course we were

oblivious, dead to the world in our beds while the temperature of the freshly pressed chenin blanc rose in the cellar. The next morning, a dawn start in the cellar was met with new wine cascading out of the tops of the stainless steel tanks and flowing across the cellar floor in a sticky, foaming river.

A nightmare of cleaning up followed, in which every hand was issued with a mop, broom or cloth and set to work. Our first priority was to reset the cooling system as the entire chenin harvest was at risk. If wine gets too warm it ferments more quickly and this will compromise its natural, delicate flavour. Luckily for us we discovered the looming disaster before the heat of the day set in and were able to save the best of that year's crop.

There are three rules that apply to the harvest. Put simply we have early mornings, hard work and late evenings. The by-products are numbing tiredness and aching bones. I manage to survive by snatching catnaps in the early evenings before serving up a hasty dinner.

During the harvest frenzy, meals become rushed affairs. Supper is often a stew placed on the stove at lunch time and left to simmer slowly and develop the hearty flavour of a well cooked casserole. Sometimes I might throw a quick pasta together late at night to relieve nagging hunger. Any job not related to the harvest is relegated to night duty.

As harvest time drags on, my to-do list and the inbox full of unanswered e-mails grow longer every day. Everything is forgotten in the drive to get the grapes into the press and the juice into tanks.

It's at this time that I resort to picking salad and vegetables by torchlight as the days fly past. At weekends when I take a walk through the vegetable gardens I discover that the curious life of vegetables has continued unabated. They constantly grow and provide food, no matter what is happening around them. Tomatoes, courgettes and beans are begging to be picked.

As the month draws to an end most of the grapes have been harvested. Only the cabernet sauvignon still ripens on the vine where it needs to stay for a couple more weeks. My time is free and the pressure eases as the wines finish fermenting and are pumped into barrels to start their secondary malolactic fermentation so they can start to mature at their own pace.

As I emerge from the cellar, the days have grown shorter and the weather cooler. I tackle weeds in the vegetable garden and harvest an abundance of produce for friends who run an organic vegetable delivery box scheme in Cape Town. I leave early the next day with a car laden with boxes of tomatoes and weighed packs of French beans. When I'm done, I'm going to treat myself to a cappucino away from the madness of harvest.

Brioche

makes 1 large loaf

350g	plain flour
10g	dried yeast
25g	sugar
pinch	salt
50g	milk, at room temperature
4	large eggs
200g	unsalted butter, softened and cut into cubes
1	extra egg yolk to use as an egg wash

Put the flour, yeast, sugar, salt and milk into the bowl of an electric mixer. Use the dough hook attachment. It is impossible to mix this by hand due to the amount of mixing required. Add the eggs 1 at a time with the mixer on a low speed. It should take about 10 minutes for the dough to become very smooth, and slightly elastic in texture. Add the softened butter, a little at a time. Continue to run the mixer for about 10 more minutes when the dough will be very smooth and glossy. Remove the dough hook, cover the bowl with a large plate and leave in a warm place for 1 hour or until the dough has doubled in volume. This will take longer in cooler temperatures. Knock back the dough by taking it out of the bowl and replacing it. Dust your hands with flour first. Leave the dough covered overnight in the fridge.
Next morning, remove the dough from the bowl and cut into 6 equal pieces, line a loaf tin with baking paper, form the pieces of dough into flattish disks and place them side by side along the length of the loaf tin. Gently brush the formed loaf with the beaten egg yolk and put to one side in a warm place for 30 minutes while you heat up the oven. Bake at 200°C or gas mark 6 for 45–50 minutes. The brioche will be a glossy dark brown on top. Leave to cool slightly before serving.

Lemon Muffins

	makes 10 muffins
150g tub	natural yoghurt
2 tubs	self-raising flour
1 tub	unrefined sugar
1 tub	melted butter
half	lemon, zested
1	large egg

Preheat the oven to 180°C or gas mark 4. Place the yoghurt in a bowl, then wash and dry the tub and use it to measure the flour, sugar and butter into the bowl. Add the lemon zest and egg, and mix all the ingredients well. Line a muffin pan with paper cases and fill them two thirds with the mix. Bake for 15 minutes. Serve warm.

Tip: Add a tablespoon of poppy seeds for a change.

Creamy Onion Tart

For the shortcrust pastry

250g	plain flour
175g	unsalted butter, cut into small cubes
1	large egg

For the filling

4	large or 6 medium onions
50g	butter
6	large eggs
300ml	single organic cream
1tsp	Dijon mustard
pinch	each of salt, pepper and nutmeg

Serves 6-8

Mix the butter into the flour until crumbly, then bind the mixture together with the egg. If the mixture seems dry, add a little cold water. Roll out the pastry and line your flan case. Let the pastry relax for 10 minutes while the oven is heating up to 180°C or gas mark 4. Blind bake for 10 minutes. Meanwhile, finely slice the onions and slow-fry them in the butter for 20 minutes or until they are soft. They should not become crisp. In a bowl, beat the eggs, add the cream, mustard, salt, pepper and nutmeg. Once your tart case is out of the oven, fill it with the fried onions, then pour over the egg mixture. Return to the oven and bake for a further 30–40 minutes at the same temperature. When cooked the flan should be golden on top and gently set with a slight wobble. Leave to cool for 15 minutes before serving.

March

The road from our farm to the nearest village, 12km away, cuts through a sea of vines, backed by mountain in every direction. It is early morning. The car is loaded with boxes of wine for delivery to Cape Town. As I stop at the crossroads to turn right into another vine-bordered road I notice the bright morning sun lifting into the sky from behind the mountain. To the right, a huge round moon still hangs in the morning sky, reluctant to disappear from the heavens.

I can already see Table Mountain in the distance. I love Cape Town, and think it is one of the most beautiful cities in the world. Today I have meetings in the city and I am anxious not to be late, but after living down a quiet country road, the congested city streets have become a challenge for me. As the main road approaches the outskirts of the city, the traffic slows to a crawl and I find myself trapped in a melee of cars that rolls slowly onward. The air is thick with fumes and there is nothing for me to do but make mental lists of chores for the farm and press on.

I reach my destination on time and find parking in one of the garages that are a part of life in this city. Meetings over, the last delivery stop is the Waterfront, where I spend time over a good cappuccino and relish the chance to stock up on good books. Having grown up in the cold climate of the northern hemisphere, I relish figs as one of the ultimate luxury fruits. Unlike me, my family don't enjoy them raw, so every year

I make fig jam, which goes particularly well with Camembert cheese. It also provides a luscious taste of summer in mid-winter. The fig is unique, as it is pollinated not by another fig, but by a fig wasp that climbs into the fruit and pollinates it from the inside. This year they are fruiting late and I am so relieved to see the fruits turning to purple at last.

I pick a basket of figs and cannot resist the temptation to eat several on my way back to the kitchen. Tonight I shall cook some with pork and add cardamom and rosemary. The juice will be splendid and I'll spoon it over mashed potatoes. I pack the rest of the figs into small trays to deliver to our friends who run the organic box scheme. I work carefully as figs do not keep well and they bruise easily. So fashionable have these plump sweet fruits become, that I could sell far more. When we ran our markets, figs would be so eagerly snatched up that I don't know why more people don't have a fig tree in their garden. I do very little to our fig trees and they never fail to produce a luscious crop.

It's been a busy Saturday, and all of us are getting jobs done on the farm. What's the saying? "A farm never sleeps." We are joined by friends and have supper out on the patio with a couple of bottles of our own crisp chardonnay. It's been a lovely evening and the witching hour is approaching as we finally get to bed. The boys fall into a dead sleep, and I can finally drift off into that pleasant, safe state of early slumber.

I am woken with the clock reading 12:30. We have been in bed for only an hour. Shock brings me to my senses as I realise what has woken me. I sniff again and then again. Fire. Looking out of the bedroom window, I see the red glow on the mountains behind us and hear that distinctive crackle as it comes to me that the mountains have been as dry as a tinderbox.

We're all up in seconds. The men fill up the tractor-pulled water sprayer. I run like a mad woman, turning on every sprinkler. Then I shift attention to our thatch roof, which is dry and vulnerable. I hose it down until it is sodden and dripping before I walk up to the road to find out what's happening. I can hear voices shouting from the mountain side, but can see nothing through the darkness and smoke.

As a neighbour's tractor comes hurtling down the mountain to refill with water, I hear that the fire started on their farm. There are no more details so I will have to wait it out. The fire seems to be under control, and mercifully the wind is not blowing. Last year a fire on these mountains burnt for days.

Danger and disastrous situations always bring out a spirit of camaraderie. Men and their machines come trundling down the mountainside an hour or so later and I don't recognise half of them. They turn out to be visitors from a neighbouring farm and after much back-slapping, chatter and promises to visit for a quick glass of wine later that day, we all make our way back to our beds. The sky is just starting to blush with pre-dawn light. For the children it has been a huge adventure. The adrenalin is flowing, and at that young age the perception of impending disaster is seen in a totally different way, without the responsibility of having to deal with consequences.

Frightening as these Cape mountain fires are, they play a vital role in the ecology of the flora for which the region is famed. Most of the fynbos depends on seasonal fires for their seeds to germinate. I finally fall into a deep sleep knowing that, despite the blaze, the charred mountainside I love so well will regenerate itself completely by next year.

The big clean-up starts when the men go off on a fire inspection. Filthy smoke-stained clothes are soaked in buckets. The pool has a thin covering of ash, as do all the window ledges. The fire is out, except for one large smouldering tree trunk, which must be kept under a watchful eye. We are all too exhausted to do much else that day.

Only an Italian could have introduced a gem like the borlotti bean to the culinary world. The beans look like bright cerise and white mottled earrings, hanging from the tripods that I built out of bamboo canes. It is almost a shame to pick them, they look so beautiful. But they make wonderful soup and dry brilliantly for storage. This morning I grab a handful for soup. It's going to be a busy day, so if I put a pot of soup on to simmer slowly I won't have to rush around thinking about supper later this evening. The advantage of having these beautiful beans growing literally outside the kitchen door is that, while they are fresh, I don't have to use dried ones, which would require overnight soaking before use.

I leave the rest of the borlotti beans to dry naturally on the vines where they grew. I'll gather them later and store them in a jar for use in winter. They add a depth and a heartiness to soups and casseroles that is most welcome. As I work, I spot a tiny pink flash of colour in the greenery. The late-fruiting raspberries are ripening, their round fruits now hang from the lush canes, inviting me to come and collect them.

These raspberry canes took about three years to establish and bear fruit. It was a long wait from the day I bought four tiny plants at a nursery and lovingly tended them, mulching them every year with copious amounts of rich compost. They love the shade of the old oak tree under which I planted them, and have started to reward me well for my trouble. Raspberries have a bit of a tendency to spread and multiply, so much so that I now have about thirty plants. I continue to nurture all the new canes that shoot up every year and religiously cut back the old growth. I will save this treasured fruit for the weekend, when all the family are home.

It is my youngest son's birthday, and as a special treat I will scatter the raspberries in the bottom of small dishes, then cover with egg custard mixture to make crème brûlée. Spooning out the rich custard and finding the contrasting flavour of the fruit at the bottom is like finding treasure.

At last the hens have started laying again. Their rhythms are affected by heat and light. Without enough light, the laying stops each winter. In summer too much heat also upsets their rhythm. We had so many eggs at the end of last year that I was giving them away. Then, the summer heat descended on us and the egg delivery system stopped. The days are getting cooler and the "old dears" are the first to start up, but I hope to get eggs from our young pullets soon. They are close to twenty weeks old, so it won't be long now. I have kept chickens for years. Even when I lived in town I always kept a few in the garden because I love to have a ready supply of fresh eggs, which taste so much better than the shop-bought ones.

Alone again, as evening falls, I sit outside my kitchen enjoying the stillness and listen to a myriad sounds. The nights are drawing in, signalling the end of summer. Close to the kitchen, there's a small dam where the frogs perform their nightly opera. There are several species that all add their own notes, but it always seems to be the big deep base of the bullfrog that starts the show.

A large Scots pine tree stands next to the dam serving as a wonderful lookout point for birds. I spy one of our large resident barn owls perched on a branch and hear it call to its mate. The volume of the frog opera increases. Then silence. They must be warning each other of impending danger as frog is a delicacy for owls. As it gets darker fruit bats come out to play, swooping from tree to tree with effortless grace. Darkness falls. It is time to turn in. I leave the nightlife to play and head for bed.

The roses have come back to life, and I am again able to pick blooms for the house. Everything in the vegetable garden looks healthier as the month draws to its close. The intense heat has dissipated. I pick baskets of lettuce, tomatoes and beans for salads. A fresh crop of mixed salad leaves pops up from the earth. I pluck the baby leaves off the plant and toss them into vinaigrette. A few days later, as if by magic, fresh salad leaves appear. It's like a false spring.

There is no pressure on my time and I am able to spend hours out in the vegetable beds. This is one of my favourite places and I'm relieved to be in the open air after spending so much time labouring in the cellar. Tall wigwams are covered with scarlet-flowered runner beans. These are not my first choice of vegetable but I love the bright lipstick-red flowers. They look so dramatic climbing like aliens up the canes.

The first apples ripen and I collect baskets full. They are sweet and crisp. My little crop of aubergines has finally turned shiny and deep purple. I cannot look at an aubergine without tasting ratatouille, so that is what we eat several times over the next few days. I love it hot or cold, served with fresh brown bread to mop up the juices. I always try to make enough so there are plenty of leftovers for lunch or lunchboxes.

Since we started buying freshly ground flour, any thoughts of wheat allergies have vanished. I am convinced that food is totally over-processed, and that the processing, not the food, is what gives everyone allergies and illnesses. It's hard to put my finger on what causes allergies, especially when everyone on the farm has a constant diet of fresh food, very little of which comes from supermarket.

The ruby red peppadews are hanging in thick clusters from the bush. I collect them, and when the basket is full I settle down after supper to thread them onto a long piece of fishing line. They will hang in the kitchen like a necklace and will dry out slowly. In the months to come I'll have a wonderful supply of chilli. In the evenings I also mix sea salt with dried lemon peel and herbs. Placed in jars this makes a superb seasoning and I fill jar after jar and store them away to use in winter and as gifts for visitors.

There's always something to put aside or preserve in some way, so friends are often in luck when they come to visit. One night a freak wind blows so many apples onto the ground that our apple supply for winter is reduced. Next morning I fill three baskets from the dew-damp grass and chop them into my largest stock pot to soften on a medium heat before being strained. Sugar goes into the strained juice which I boil in batches with various flavourings. By afternoon I have bottled chilli jelly, mixed with berries for colour, mint jelly, lavender jelly, and my favourite: lemon grass and ginger. I fill a cupboard shelf with the small jars, content that the windfall apples have not gone to waste and that there is abundant stock of little gifts to present to friends in the coming holidays.

Easter is approaching and on a weekend home from school my boys ask me for pancakes. I make batch after batch, which they devour with lemon and cinnamon. There's another batch for dessert later that night and we roll them into cones filled with ice cream, a firm family favourite.

With the evenings starting to cool and being in the holiday mood, I'm thinking of chocolate treats. We have a crowd of children joining us for the holidays so plenty of food and treats are needed. I spend an evening stirring and melting chocolate in the double boiler. The evening is chilly, and I am forced to close the French doors. Kitchen tiles are cold under my bare feet, summer is fading fast.

Borlotti Bean Soup

For the soup

4	rashers of bacon
50ml	olive oil
1	large onion, finely diced
4	cloves garlic
2	large potatoes, peeled and cubed
4	tomatoes, peeled
400g	borlotti beans
1½ litres	water
	salt and pepper to taste

For the bouquet garni

2	sprigs sage
1	sprig rosemary
2	bay leaves

Tie these together with string or form a pouch from a piece of muslin and tie closed.

Serves 4-6   Cut the bacon into small lardons and fry gently in the olive oil. Add the finely diced onion, garlic, potato and lastly the tomato. Add the beans, bouquet garni and the water.
Bring the pot to the boil, and then let the soup simmer for 90 minutes on a very low heat.
Remove the bouquet garni and season to taste. Allow to stand overnight as this soup always tastes best the next day.

Tip: If you are using fresh beans there is no need to soak them first, but if using dry beans you must. Season with salt only at the end of cooking or else the beans will be tough.

Crème Brûlée
with Raspberries

500ml	single cream
8	egg yolks
75g	sugar, plus a little extra for making the brûlée
2 drops	pure vanilla essence
24	raspberries
6	small ramekins or ovenproof bowls

Serves 6

Preheat the oven to 140°C or gas mark 1. Heat the cream in a pot, but don't let it come to the boil. Beat the egg yolks with the sugar until pale and fluffy, add the vanilla. Pour the hot cream into the egg mixture, stirring gently. Place four raspberries into each of the ramekins and cover with the egg mixture. The brûlée should only be about 3cm thick.

Place the bowls into a deep baking tin, and fill with hot water until the water reaches halfway up the ramekins. Bake for 25 to 30 minutes. The custard should have a wobble when touched. It is very important not to overcook or the custard will split. Leave to cool before placing them in the fridge until you are ready to serve. Before serving, sprinkle a slight layer of caster sugar over the top of each dessert, and caramelise using a blow torch – a grill really does not work.

Loin of Pork with Figs

1.5kg	boned pork loin
	salt and pepper
10ml	olive oil
6	figs
200ml	white wine
6	crushed cardamom pods
1 tbsp	chopped rosemary

Serves 6

Score the fat on the loin with a sharp knife, slicing through the fat gently at 1cm intervals and rub salt and pepper into the skin. Drizzle a little oil over the loin and place skin side down in a roasting tray that has been heated up. Pop this into a hot oven at 180°C or gas mark 4 for 45 minutes, then turn the meat over and return to a slightly hotter oven at 200°C or gas mark 6 for a further 30 minutes. If the tray has dried out, add a couple of teaspoons of water to prevent the meat from burning. The hotter heat will assist in crisping up the skin.

Cut the figs with a criss-cross on top and place them around the pork. Return to the oven for ten minutes. Jab a knife into the centre of the joint. The juices should run clear, and the pork should be cooked by now. Remove the pork and figs from the tray and put to one side to rest. While it is resting, deglaze the tray with the white wine, then add the crushed cardamom and chopped rosemary, stir together and bring to the boil to reduce. Slice the pork and serve with the figs and reduced juices poured over the top. I always make a mashed potato laced with chopped sage to go with the pork.

Spelt & Honey Choc Chip Cookies

200g	spelt flour
75g	spelt flakes
180g	unsalted butter, at room temperature, cut into small cubes
1 tbsp	honey
100g	unrefined white sugar
50g	dark chocolate chopped into small pieces

Place all the ingredients into a bowl and mix well.

Roll the mixture into small balls and press down gently with a fork. Place the disks onto a baking sheet or greased and floured baking tray and bake for 8–10 minutes in a pre-heated oven at 180°C or gas mark 4. The biscuits will be soft when removed from the oven. Let them firm up and cool for about 5 minutes before placing them on a rack to cool.

Tip: Spelt flour and flakes are available at health food stores and have a wonderful nutty flavour. If you cannot get them, replace with plain flour and oat flakes.

April

A couple of hens have gone missing. I spend a morning searching for them, certain that they must be scratching about somewhere. I hope someone has not stolen them. There is no sign of scattered feathers anywhere, so I doubt they've been kidnapped by a rogue dog. For the next couple of days I keep a close eye on the chicken shed. They never return and we find no evidence of their fate, such are the mysteries of farm life.

I take my last swim for the season as the water is getting cold. It is refreshing but I whiz around the pool and get out quickly. I must think about covering it up for winter after the school holidays. Easter school holidays are my favourite. They hold so many memories of when the children were small.

In those days I would set my alarm early on Easter morning so I'd have time to hide little chocolate eggs all over the garden. The children would wake up and fly out on a search mission. One year the dogs spoilt the game because they got to the eggs first. All that was left were shiny fragments of bright coloured foil papers fluttering in the flower beds. The place looked like Christmas at Easter. We managed to rescue the few eggs that the dogs had missed and the children gobbled them down, fearful to let them out of their sight. Now the boys are older, although we still have eggs, we no longer have to hide them. The harvest is over for another year and all the grapes are crushed. The vines are tinged with gold, which will soon turn to vermillion. The wine is fermenting and maturing in tanks and barrels. The heady, fruity scent of fermentation has faded, followed by the more subtle aromas that are developing. I keep a careful eye on the cellar, but the pressure is off.

Now we can relax and enjoy the last of the summer sun, while we wait for our satsumas to ripen. The satsuma is a type of mandarin orange, which originated in China. Buddhist monks, attracted by the deep golden colour of the fruit that resembled their robes, introduced them into the Satsuma Province of Japan, where they flourished. Today they are a popular autumn fruit in many areas around the world. I love them. Beneath that leathery skin is a sweeter version of the ubiquitous orange. Satsumas also have the advantage of being pip-less, making them easy to eat. They are guzzled down as soon as they ripen, providing the family with a good vitamin C boost before winter. All of the damaged fruit is squeezed to make jugs of sweet fresh juice.

Back from a couple of days at the beach, swimming in the cold ocean to wash away the summer dust, I plan an Easter lunch. I'm making a large garlic-infused leg of lamb with roasted vegetables. We spend a lovely long weekend lazing around, enjoying our farm. We sleep late in the mornings and spend time together, with no deadlines to meet. We indulge in a glass of wine with lunch, followed by an afternoon stroll and loud games of French cricket in the garden.

I bake a cake for afternoon tea, choosing a rich classic Victoria sponge, which is a firm favourite. This one is laced with lavender and topped with a tangy lemon icing that cuts through the richness of the sponge. After four days of this lazy, blissful life, we feel revived and ready to tackle the cold winter months that lie ahead.

When the house is full, I bake bread several times a week, less often when the boys are at school. I usually make it first thing in the morning. Once you start baking your own bread it becomes addictive, as the end product is amazing. Bread-making became a habit after I had bought an emergency supermarket loaf, years ago. It listed 14 ingredients on the packet, some I had never heard of, and some I did not wish to eat. I thought that 14 ingredients to make a basic food that should comprise flour, yeast and water was a bit much.

The wonderful smell of baking bread draws everyone into the kitchen. I've noticed that fresh bread is a magnet for my youngest son, who eagerly attacks a fresh loaf even before it has cooled. As a result, my kitchen counter always seems to be covered in crumbs and our Jack Russel follows him around like a vacuum cleaner. We often toast day-old bread for lunch and give it a good rub with garlic and a drizzle of olive oil, and I lay out a selection of toppings like sliced tomato and creamy goat's cheese.

Recently, I have been baking bread with spelt flour, which has less gluten and is so much healthier. Spelt is an ancient grain, widely used by the Romans, but it's not as popular today, as it produces a much lower yield than our modern wheat. But the fact remains that it is much better for you and has a distinctive nutty taste.

We harvest the olives after the Easter break. It's not a large crop as most of the trees are quite young, but how I look forward to producing our own olive oil in a few years' time. For now I must be satisfied with simply brining our olives. We harvest fifty kilos, which take up residence in huge buckets of water in the feed store. Every few days, for the next three or four weeks, this water has to be renewed. Draining the water and refilling the buckets is a laborious job, but essential for the olives to lose their intense natural bitterness. I add large amounts of sea salt to speed up the process. After a few weeks the bitterness will have seeped away and the olives can be stored in large preserving jars, filled with a weak vinegar solution and infused with fresh sprigs of herbs, dried fennel seeds and peppercorns.

April is marked on my calendar as the "Drying" month. Herbs have always been essential to my cooking. I love their taste and they are so full of benefits. Ah, the magic and mystery of herbs. People have been burnt at the stake for using them. In more enlightened times they are used to save countless lives as many modern medicines are derived from herbs.

I grow an enormous selection of herbs in the garden and, as many will die down in the colder winter months, I set about drying them for future use. My method is simple. I spread them out on trays and put them out in the autumn sun to dry. If the weather looks dodgy I bring them inside to the warmth of the kitchen. Most herbs dry in a couple of weeks, but the woodier ones like rosemary take longer. Thyme, sage and oregano all dry exceptionally well, but I think the sage, *salvia officinalis,* holds it flavour the best. What a herb. Wise people are called "sage" for good reason. It is said to promote clear thinking and was much loved by the ancient Greeks, valued by the Chinese and deemed a sacred herb by the Romans. That legacy lives on; apparently it is being tested by today's scientists as a possible treatment for Alzheimer's disease.

The Italians are the masters of cooking with sage, a skill they inherited from the Romans. Anyone who has eaten a good plate of Italian pasta with sage butter sauce will agree. I have never tried to recreate the sage ravioli that we once ate in a rural Italian trattoria but I remember it as an ambrosial moment. My ultimate comfort food is sage risotto, simple but deliciously creamy rice infused with the aroma of this invaluable herb.

Today I am snipping off dried fennel heads, each an umbrella of seeds. Some plants stand head high and the smell is intoxicating. The seeds are almost dry, so after a few days on the tray I will settle down one evening, and carefully pluck them off. Some I will re-sow to grow more and I might just get another crop before winter. Others will go into a jar to be used to add flavour to an assortment of chicken dishes.

I move on to my beloved lemon verbena. The leaves have a strong lemon sherbet scent and I use them for tea. Fresh leaves do me in summer, but now I am collecting the last of them for my winter store. This plant hates winter and the leaves will fall off if not collected. I have trained each of my four plants to grow like topiaries. Each with their long trunks supported in the centre of four square herb gardens. Each square is divided into four triangles and each triangle holds a different herb.

The soil is very fertile and I suspect that in years gone by the area was used as a rubbish dump. When we first moved here I kept digging up pieces of shattered old china, many of which I have kept in a box. The nearby town of Tulbagh was damaged in an earthquake measuring 6.3 on the Richter scale in 1969. A few old buildings on the farm collapsed during this seismic event, which shook the area late at night, leaving eight people dead. Perhaps the china shards were among the damaged possessions that might have been buried here. How I would love to know.

Today's verbena tea is beautifully scented; I mix a handful of leaves with a spoon of honey, pour over boiling water and leave it for three minutes to infuse. Taken before bed I am guaranteed a peaceful sleep. Autumn is a very good time to take herb cuttings. Some of my lavender and rosemary plants are eight years old and getting a bit woody, so we cut them back into balls. They look lovely as rows of green-grey balls. It is painstaking work as we have hundreds of plants. I plant the tender tips into plant trays filled with our own rich compost, as my trusty garden helper clips away. The small cuttings will stay in trays until next spring when they'll be transferred into larger pots, and eventually to the land. We've propagated thousands of plants this way, which is something I find so rewarding and sustainable.

I dried the last of my beautiful purple figs in the oven last week, leaving them in there for a day with the oven on its lowest setting. I left them a little soft as today I will mix them into a large batch of granola. I never seem to be able to make enough. It's so easy and, topped with yoghurt and fruit, makes a healthy and filling breakfast. And it disappears faster than I can make it.

We grow two varieties of apples, the nicest being the Cripps Pink. These apples are the most vibrant pink and are incredibly sweet and crunchy. This year the trees are covered and each day I pick some. By evening they are all eaten, the windfalls feed the pigs and horses, and other bruised fruit is cooked and stored in the freezer for future apple tarts and crumbles. Tonight I shall make a spiced apple crumble, adding a few hanepoot grapes. The last few grapes still hang on the vines amongst leaves that are now turning several glorious shades of yellow, orange and red.

The colours of autumn are a triumph of nature. It is morning and a light mist lifts to reveal a patchwork of colour. As I walk down to the citrus orchard my shoes become sodden from the heavy dew on the grass. I stop at the river paddock to stroke the horses that are full of energy. They canter around me in the paddock leaving heavy footprints on the damp grass. The two little Welsh ponies come up to nudge me in the hope that I have a treat for them.

The boys like to harness them to a cart, and off they all go flying around the vineyards. The ponies love it too. "Welshies" are wonderfully hardy little ponies. Originating in Wales, they have roamed the Welsh mountains since 1600 BC. When Julius Caesar invaded Britain he was so impressed with the tough little ponies he took some back to Rome. Thus started their journey around the world.

As a child I used to ride a "Welshie" and remember well the day he threw me while riding through a spooky wood. We were cantering through an area of whispering trees when I suddenly came off, landing on the earth with a thud. Dizzy and disorientated, I climbed right back on and slumped over his shoulders. That trusty little pony simply walked me right up to our garden gate. I shall never forget the shocked look on my mother's face as she opened it. I have had great respect for these sturdy little ponies ever since.

Walking on through the orchard I pluck a satsuma from one of the trees, peel it and pop a segment of fruit into my mouth. It's juicy and delicious but a little too acidic and not quite ready to harvest.

The tomato season is drawing to an end. Crouched on my haunches, I work my way along the rows of tomatoes, picking several large basketfuls. The rest of the day is spent roasting them in trays with herbs, salt and a splash of olive oil. Then I pack them for the freezer, they will be used in coming months to flavour stews and sauces. Later we need to move inside halfway through supper since the rain arrives in torrents. The sky lights up and the earth shakes under a clash of thunder. I hastily run out to gather washing from the line, and return soaked. The earth drinks thirstily as we have had only a few little showers since December with plenty of wind to dry the soil.

Gullies fill up, rivers flow faster and we see our first puddles in four months. The smell is heavenly; everything is getting a good wash. The rain lasts for three full days. The ground is sodden and strewn with leaves, which are gathered up and added to the compost heap. This storm reminds me that summer has passsed, and I start to gather together remnants of our outdoor life.

Baskets full of dry oak leaves are scooped from the pool and consigned to the compost heap, and then the pool is covered for winter. Into the attic go the boys' pool toys and I move my little green writing table from under the pin oaks up to a sheltered patio. That night I light our first log fire to take the edge off the evening chill.

Now I pass the evenings writing. This table, having belonged to a glamorous aunt, holds great memories. It was the kitchen table in her Chelsea flat for years. I had it shipped out here after her death and it serves as a constant memory of her dynamic life spent travelling and working in the film industry.

Not being a great lover of television, I am happy to sit and write for hours during the winter months. Suddenly the evenings seem a bit darker and much more gloomy. The sun still sets at roughly the same time as a couple of days ago, but the feeling of winter is heavy in the evenings that follow. A jury of grey clouds hanging low in the sky passes judgement on the season to come.

Spelt Bread

	makes 1 large loaf
250g	plain flour
250g	spelt flour
10g	yeast
300ml	water
pinch	salt

Pour the 2 flours into a bowl, then add the yeast, water and salt. Mix together using
the dough hook attachment to knead the bread. Knead for 3 minutes by machine and
10 minutes by hand. Put in a warm place covered with a plate and leave to rise until
the dough has doubled in size. Now preheat the oven to 190°C or gas mark 5.
Knock back the dough and place onto a floured baking tray, in whatever shape you desire.
I usually make an oval. Leave to rest for a further 15 minutes. Bake for 30–35 minutes.
The bread should be pale golden brown and sound hollow when you give it a knock.

Lavender Sponge Cake

250g	unsalted butter
250g	unrefined sugar
4	large eggs
250g	self-raising flour, sifted
10	fresh lavender flowers, finely chopped

Preheat the oven to 180°C or gas mark 4. Beat the butter and sugar until pale and creamy, add the eggs 1 at a time then gently fold in your sifted flour and chopped lavender flowers with a spoon. Place into a well-buttered cake tin and bake for 30–35 minutes.
A skewer should come out clean with no traces of dough on it. Leave the cake to rest for a few minutes before removing it from the tin.

Tip: You can also make small cakes in a muffin mould, bake for only 20–25 minutes.

Granola

fills 1 large storage jar

300g	organic rolled oats
150g	mixed seeds
200g	chopped nuts of your choice
150g	runny honey
100g	dried fruit (figs, sultanas, dates or apricots)

Preheat the oven to 150°C or gas mark 3. Line a baking tray with baking paper. Mix the oats, seeds and nuts together and pour onto the paper, drizzle over the honey. Loosely mix the honey into the dry ingredients. Bake for 10 minutes. Remove from the oven, sprinkle over the chopped dried fruit, mix in roughly and return to the oven for a further 5–8 minutes.
Let the mix cool totally on the tray before storing in a sealed jar.
Keeps for 3 or 4 weeks.

Sage Risotto

50ml	olive oil
6	slices streaky bacon, chopped into lardons
2	cloves garlic, crushed
250g	risotto rice
1 litre	hot chicken stock
handful	fresh sage, chopped
50g	butter
50g	Parmesan cheese, grated
	salt and pepper to taste

Using a large pan fry the bacon in the olive oil, add the crushed garlic and rice, gently mix the rice until it is coated with the oil. Start ladling in the hot chicken stock, stirring continually and adding when needed. Repeat, and after about 20 minutes the stock should all have been absorbed into the rice and the mixture should be creamy in appearance. Turn off the heat and quickly stir in the sage, butter then cheese. Season to taste and serve immediately.

May

As they hang on the trees, the satsumas resemble bright orange Chinese lanterns. The cold nights have worked their magic, turning the green-tinged skins to bright orange. The fruit is juicy and sweet, and I eat several each day. We start the picking late each morning for as long as the weather holds out. In the early morning the fruit is covered in dew and it will spoil if collected while wet. The kitchen is full of satsumas stored in baskets and I have fun inventing ways of cooking with them. The children take bags of the ripe fruit to share with friends at school. These bags are bursting at the seams. I deliver a load to our local crèche.

The newest addition to my culinary repertoire this weekend is a satsuma cinnamon honey cake, the scent of spice and orange can be smelt throughout the house. I make batches of marmalade for a friend to sell in his farm shop and the huge copper pot bubbles and simmers on top of the stove while it thickens. Now the search is on for jars with matching tops to sterilise in the oven.

Unable to have my morning mug of coffee in the garden, I sit at the large kitchen table and plan my day. I cannot function without lists. I forget things if I don't have written reminders, and as I march further into my 50s, the problem does not seem to improve.

Coriander is one of the few herbs that grows well in the cool damp autumn weather, the others being chervil and rocket. All three grow quickly, so today I am planting out several rows of each, and a row of lettuce too. I removed my bean tripods last week and the vegetable beds are looking so bare.

This is the only time of year that I can successfully grow rocket with some degree of success. The cabbage butterflies have disappeared for the winter. Their larvae love nothing better than a crop of rocket and strip every tender young leaf they can find. I've never been able to find a solution to this problem. I have tried to disguise the smell by interplanting the rocket seeds with stronger smelling plants. I have tried planting a sacrificial crop of nasturtiums, with no success. The butterflies will not be dissuaded from my rocket. They lay thousands of eggs, which quickly hatch into brightly coloured caterpillars that feast so greedily they can strip the plants within hours. I've even tried picking them off by hand, but I cannot keep up with this army.

I love broad beans, and now we are sowing thousands of seeds, saved and dried from last year's crop. These are planted between the vines, as they supply a rich source of nitrogen to the soil. They grow while the vine is dormant so are perfect companions to the vine. Their culinary value is an added bonus.

The compost heap at the bottom of the vegetable garden is full of rich dark compost. I am shovelling spade-loads into seed trays into which I pop sweet pea seeds. Last year's yield of this fragrant flower was poor and I

intend to rectify this for next spring. All the kitchen scraps that don't go to the pigs and chickens are piled onto the heap. To this I add the odd newspaper, grass cuttings, some muck from the chicken shed, and in winter, the ashes from the hearth. After three to four months, with the odd turn, it produces wonderful compost. We have huge heaps at the far end of the farm to mulch the vines. The little heap close to the house is used for the vegetable gardens.

The compost does its thing, turning to black gold in a clever structure built by the boys one school holiday. This consists of a square with two old vine poles driven into the ground at each corner, slotted between these corner poles are more old vine poles, which form the walls. The structure has three walls, leaving the front open for dumping and spading out the fresh compost. I have two of these alongside each other so I can use the compost from one while making new compost in the other. It works like a dream.

Two old trees stand at the back of the farmhouse. On the one side is the old lemon tree that provides a steady supply of huge lemons with nice thin skins. On the other side is a large avocado that also provides it's share. The trees do double duty as the retainers of the family washing line, which is strung between them. The avocado has also done duty as companion to two small boys who used it as a giant crane for hoisting things up into the

Land of Play. Both trees are laden with fruit. I am the only member of the family who enjoys avocados so I feast for two months of the year and give the excess away to staff and friends. Tonight, after a sunny autumn day, my solitary evening meal of guacamole with wholewheat toast is supplied by these two trees. I sit in the kitchen reading with the window wide open, listening to the owls court each other in the age-old oak trees. I love every mouthful.

Tiny oak trees pop up in the flower and vegetable beds. The trees surrounding our house have dropped their acorns all over the garden. Many perish, the others we transplant, careful not to damage their long tap root. European oak trees stand next to most of the old Cape Dutch houses. They were planted to provide a source of wood for making doors, windows and floors. It was only later realised that the warmer climate in South Africa makes the trees grow too fast, and the oak is too soft for fittings and furniture. Our trees are giants and must be close to two hundred years old. They provide our dappled shade in summer, but a couple of them have become quite hollow inside and I fear a winter storm may bring one of them tumbling down.

When my children were younger they would feed fallen acorns to the pigs. Some years back we all went off to visit a cheesemaker in the pretty little town of Montagu, which sits halfway between our valley and the Karoo. While we were looking around the cheese maturation room, my youngest son disappeared outside. He told us afterwards that he'd gone to investigate the squealing that he'd heard outside. He soon came running back, delighted to report that the field behind the cheese shed was alive with little piglets. They were happily living off a regular supply of whey from the cheesemaker. After just a little persuasion, four dear little piglets were selected to become the first pigs to take up residence on our farm.

A few months later they were joined by the "Grand old Duke of Pork". This fine-looking young boar was given to us by a friend in exchange for a couple of cases of wine. Since then several litters have been born on the farm, supplying us with delicious home-grown pork. The pigs pass their days running and snuffling around in front of the stables. They live on a natural diet, supplemented with acorns, vegetable scraps, excess fruit and whey. Free-range pork is not at all like the pale commercially produced product available in supermarkets. The meat is much darker and has less fat, due to the fact the pigs are outside in the sunshine and are free to run around all day enjoying their lives.

My favourite way to cook pork is to make sausages, and one of my finest investments has been a small sausage-making attachment that fits onto my food mixer.

After mixing pork mince with garlic, herbs and a generous sprinkle of salt, the filling is fed into sausage casings that I order from the butcher. My younger son is a master at cooking the sausages in a frying pan on top of the stove. No sooner have I tied up the ends of the sausages than they hit the pan at a sizzle. I think that the best accompaniment for the sausages is a winter cabbage and apple salad. This crisp salad is a perfect match for the rich sausages. But my sons have other ideas and the sausages are often hijacked straight from the pan and dipped into a waiting bowl of tomato sauce.

As the season changes, there are fewer flowers in the garden. These days I tend to pick big bunches of lavender for the house. The hardy Spanish *lavandula dentata* variety flowers all year round, ensuring that I have flowers in the house even in the winter. Sometimes I snip off the flowers to use in a recipe, and I've used them in cakes, biscuits, ice cream, preserves, and with grilled meat. Tonight I shall grill some pork chops on an open fire with lavender and sage. We will slice up some day-old bread, drizzle it with olive oil and grill it while we wait for the meat to cook. The smell of the meat grilling with herbs is a real scent of autumn.

The summer crop of *lavendula angustifolia* is made into oil. Lavender has incredible medicinal properties; I use the oil in my evening bath to ensure a good night's sleep.

I also make a cream that the whole family uses for cuts, bites and sunburn. Lavender is a natural antiseptic so the cream is a superb healing lotion. One of the loveliest sights I have ever seen or smelt is the lavender harvest in the South of France. Field after field of lavender is harvested by custom-made machines, and the landscape is a purple haze that you can smell from miles away.

This inspired me to plant a stream of lavender between the natural river bank and the orchards. It took a week to propagate thousands of small cuttings, putting them into trays filled with sand and compost. During that week my hands smelled of lavender and I lived, slept and dreamt of lavender. The following year the cuttings were planted out and now, two years later, I can enjoy my own purple haze once a year.

The swallows have left. Their nests in the barn are empty, and the place is desolate without them swooping in and out.

The last few apples still hang on the tree. It's a sad time of year. When the apples and satsumas are finished, we will have no fresh fruit until the first strawberries ripen. Until then, I have to rely on preserved and frozen fruits, supplemented with organic shop-bought bananas. The large chest freezer in the storeroom is full and the kitchen shelves are buckling under the weight of jars.

Eating produce in its season is so important. Our bodies need to live in tune with the

seasons as part of the natural cycle. People have moved away from the centuries-old way of living in communion with the land and what it provides for us.

Organic farming has been an extremely fascinating journey. It has taught me respect for the earth and a realisation that patience is a vital part of life. Mother Nature respects no boundaries and can create turmoil that will upset the best-laid plans. To survive, we all need to return to living in harmony with nature, to bend with the wind and to live with the other creatures, great and small, that call this planet home. I shudder to think how many insects and microorganisms have been made extinct, by man's greed and need to control. The Roman philosopher Horace once opined that "a man with greed will always be a man wanting".

Our red wines are ready, filling the cellar with the scent of berries as I fill the last of the cabernet sauvignon barrels. These will be the new home for this rich deep red liquid for at least the next 18 months. We will have to monitor its development and judge the right moment by taste. The rows of barrels sit in a silent order, and I love to run my hands over the smooth oak curves of these beautiful containers.

The white chardonnay barrels are already full of this year's wine and I check them every week, lifting the bung to smell the evolving wine inside. Then I top them up with wine from a tank to replace the liquid lost through evaporation.

The wheat farmers of the area have just planted their crops, so I decide to do some research of my own. I dig over a small patch of bare earth denuded of summer producing crops. Then I divide it up and plant out a bag of my beloved spelt seed.

Spelt is apparently much better for you than conventional wheat, and I intend to see if I can grow it. At the very least it will condition the soil over winter. I also have a packet of flax seed, to sow as well. Although I fear it might be a little early for flax, I love their blue flowers and hope that these will not perish in the cold winter months. Next I scatter some poppy seeds for colour, and sweet pea seedlings go into the ground along the long wall. By the time they come up I will have installed a lattice of string across the wall to help them climb up. When spring arrives the garden will be filled with their intoxicating scent.

I've been crouching for hours, replanting strawberry runners, and my back aches as I stand. But I have added a whole extra row to the strawberry patch. At the end of summer the mother plants send out small runners which will set root and grow into new plants. I have carefully detached these and planted them into a new row. By summer they will be small new independent plants and we can look forward to many more strawberries.

Next to this plot is a patch of lemon thyme that's getting a little old and woody, so I start to layer it by piling soil over some of the longer stems. This will regenerate new roots and growth. I can cut the woody old stems right back down to the ground. The earth is very cold and my hands feel numb. Time for gardening gloves, which I hate, but my poor hands are looking shoddy. There is so much dead growth to be cut back in the flower borders so I set to work with the secateurs. When I've finished, the once-vibrant cutting garden looks decidedly forlorn. Grabbing a cup of tea I sit on the grass and inspect my hard work. There is plenty more to do.

Two new lambs are born. What a joy, they look so happy and full of life bouncing around the almond orchard, jumping high in the air. I can't wait for my youngest to see this. I edge closer to take a look but they're alerted by my footsteps and flee, bleating, to hide behind their mother. I sit quietly hoping they will come closer. JJ, the Jack Russell senses something and comes bounding over the gate to find me and all the sheep scatter. My poor dog looks at me with alarm. I cannot be cross with him, especially as he does not go chasing after them.

What I notice is a lot of dead wood hanging on the almond trees. Tomorrow I will muster some help and clean the trees up, before the blossoms start to form. These fine old trees need a little pampering. We work on the orchard all day, lopping the dead wood and gathering it up for winter fires.

The sky is clear and the sun is warm. Instinctively I know it is too warm for this time of year. A breeze picks up from the north and I know a cold front will follow soon. I return to the kitchen late in the afternoon to find it full of leaves. I've left the French doors open, and the wind has delivered hundreds of oak leaves into the house.

The weather is wicked. Cold rain lashes the landscape. I find it hard to believe that only yesterday I was able to work in a T-shirt. The fire is started with a few small almond twigs and I close the shutters to stop the rain flooding through the doors. The lights flicker off and on as the house is suddenly shrouded in darkness. My son sends me a text to tell me that the weekend hockey match is cancelled. This evening I make myself a bowl of parsley and leek soup. This is girl food, but could also be called medicine as both leeks and parsley are so full of vitamins. I sauté these two with some garlic, then add it to a boiled potato together with a little of the potato water. More fresh parsley is added, then it is blitzed, given a slosh of cream, reheated and "voila", I pour vivid green delicious soup into a bowl. I curl up on the couch in front of the fire to savour this nourishing meal.

Tomorrow the house will be full again and before I snuggle into bed, I stop to take two chickens out of the freezer.

Satsuma, Cinnamon, & Honey Cake

250g	unsalted butter
250g	honey
50g	unrefined white sugar
50g	brown sugar
1 tsp	ground cinnamon
1	satsuma or a small orange, zested
4	large eggs
30g	self-raising flour

Preheat the oven to 180°C or gas mark 4. Place the butter, honey, sugar, cinnamon and orange zest in a pan and gently heat until the sugar has dissolved. Add the eggs 1 at a time and, lastly, fold in the flour. Pour the mixture into a well-buttered cake or loaf tin. Bake for 45–50 minutes. Cool for a few minutes before removing from the tin.

Pork Chops Grilled with Lavender & Sage

1 tbsp	lavender flowers, chopped
1 tbsp	sage leaves, chopped
pinch	salt and pepper
50ml	olive oil
6	pork chops

Serves 6   Mix the chopped herbs, salt and pepper, and the olive oil. Rub over the chops and leave to one side for 1 hour for the flavours to infuse.

Grill on a grid above an open fire for 4–5 minutes, depending on the thickness of your chops. Turn the chops and cook for a further 4 minutes.

Guacamole

3	avocados, flesh removed and mashed roughly with a fork
1	small onion, very finely chopped
1	small mild chilli, finely chopped
2	cloves garlic, finely chopped
3 tbsp	olive oil
	salt and pepper to taste

Serves 6   Mix all the prepared ingredients in a bowl with a fork. Spoon over the oil, add salt and pepper to taste. Serve on slices of wholewheat toast.

Homemade Sausages with Apple Salad

makes about 12 generous sausages

1kg	pork mince
5g	salt
4	cloves garlic, crushed
1 tbsp	each of sage, thyme, rosemary and oregano, finely chopped

Mix all the ingredients together and carefully fill the sausage casings. Don't go too fast or your casings will split. Cook the sausages for 15–20 minutes over a medium heat in a frying pan on top of the stove. Use a splash of olive oil and turn regularly.

Apple Salad

1	apple
1	red baby cabbage
1	white baby cabbage
1	small onion
1 tbsp	wholegrain mustard
100ml	olive oil
	salt and pepper

Finely slice the apple, cabbage and onions. I like to leave the skin on the apples, but remove the cores. Mix in the mustard, pour over the olive oil and add salt and pepper to taste.

June

My growing boys never stop eating. And I never stop watching what I eat. I pulled a lovely Adelle Davis quote from a magazine last week, which I thought was rather apt. It is now stuck next to my computer, and it pretty much sums up how I feel about food. It says, "We are indeed much more than what we eat, but what we eat nevertheless helps us to be much more than what we are."

June tends to be a bit of a nothing month on the farm. It's too late for fruit and too early for blossoms. The sole exception is my beloved lemon tree that stands outside the kitchen door. At this moment, it is laden with huge juicy fruits that beam through the dull winter days like the lights on a beacon. This versatile fruit originated in Asia and is now grown throughout the world. In colder climes this crop is in such demand that the trees are planted in greenhouses. In Italy lemon trees are often found planted in huge pots so they can be moved to sheltered spots in the winter months.

This sunshine-yellow fruit is quite incredible, simply bursting with vitamin C, and it has many other benefits. If I cut myself in the kitchen or out in the garden, I reach for a lemon with its great antiseptic qualities. I will often use lemons to clean stubborn stains or deal with smells from the fridge. It's still a mystery to me why most commercial growers insist on waxing lemons, as this prohibits the use of the wonderful flavours extracted from the rind. I flavour cakes and biscuits, puddings and sauces with lemon zest. I use the juice for marinades and dressing and to deglaze a roasting tray for chicken or lamb, where a good squeeze of lemon juice will add a great fillip to the flavour.

Given the choice of just one tree to plant, mine would undoubtedly be the lemon. Fortunately, we have the space for more than one tree and we've planted ten along one of the farm roads. We have also planted limes and I look forward to the time when I can introduce this fragrant flavour to our meals. A day rarely passes when I don't pick a lemon from the tree. My mother freezes slices of lemon very successfully, and a friend of hers freezes slices in ice cubes for her G&T.

As dusk falls, I collect the last few cherry tomatoes for myself, filling a colander with them. The vegetables that I grow on the farm are shared with the staff, but these I will keep for myself. Chopped fine and mixed with diced onion, some coriander and a splash of olive oil, this simple meal will be a tribute to the final tomato harvest of the year. For the next five months, I will be forced to live without fresh tomatoes. Maybe it's time to look at installing a small tunnel.

The sun is setting as I stroll back to the kitchen. The sky is blushed with the colour of a ripening tomato, and guinea fowl run helter skelter in front of my path, finally taking flight into the oak trees for the night. On most mornings, the valley fills up with thick mist, which later clears to reveal a fresh green landscape. The chill lifts and the short sunny days are beautiful. The light is soft as it warms up the land.

I walk down to the riverside paddocks with four of our horses. As winter approaches we overnight them up at the paddocks close to the house. It's warmer there and they seem to prefer it. This came about when, one chilly misty evening, I was turning in for bed and heard footsteps outside the kitchen. I was alone, except for my sons who were much younger then, so gingerly opened the door and was astounded to see one of our Welsh ponies staring me in the face.

Behind him stood "Stompie", our grey Percheron. Their visit had required a fair amount of dexterity, as they'd walked through the vineyards from the river, then down three steps and along a garden path to reach the kitchen door. It struck me that here were two sentient beings that had made a plan to tell me how cold and uncomfortable they were down in the riverside paddock, and that they needed to be moved to a warmer place at this time of year.

I stepped into the cold night air and walked them up to the top paddock next to the house; no need for halters. We just walked together, and once inside the paddock they got down to the business of grazing on the wet grass. Ever since, we ensure that the horses are brought to the top paddocks at night.

Today, walking down to the river with them, I notice large amounts of mushrooms growing in a circle. I make a mental note that I must do something to improve my knowledge of mushrooms. We're all fond of these delicious edible fungi, and how wonderful it would be to collect our own on the farm. As if by telepathy a friend arrives that evening with a brown paper bag stuffed with beautiful cultivated shiitake mushrooms. This gift is a real treat. When I was a child my favourite supper was always a simple dish of sautéed garlic mushrooms on a chunk of toasted baguette. To this day, when I am on my own, I revert to this simple supper. Tonight I think I must make something a little more substantial so I settle for sautéing the mushrooms in the same way, to be served on a platter of tagliatelle with a dollop of crème fraiche and chopped sage. My friend stays on for supper and we wash the feast down with a glass of red wine.

Recently, when a dear friend underwent a double heart bypass, the surgeons suggested he drink a glass of red wine a day to maintain this piece of internal machinery. Now there, I thought, is a sensible doctor. Red wine contains resveratrol, proven by scientists to help prevent inflammation and blood clots from forming in the heart. Any Frenchman, as he knocks back his daily lunchtime carafe of red medicine, will tell you that it's obvious and he has known this for years.

June is when we do the structural work in the garden as many of the vegetable beds are bare. With my two sons Nic and Sam as assistants, I have in past years done some excellent development work. Together we have laid out gravel pathways lined with bricks. We have built a rose garden and even hijacked the digger loader to turn a dodgy grassy area into a scented border.

We have even made a nursery area for propagating young plants. The boys are rewarded for their work with extra pocket money. I am rewarded with the pleasure of a well laid out garden in an area that was just mud and grass when we moved here. As with most things in life, it is a work in progress, and each year I think of a new project. This year's plans include a vine-clad arbour, for which I am trying to drum up enthusiastic support and labour.

The coriander and rocket that I planted at the beginning of last month have shot up. The chervil seeds I scattered in the old half wine barrel are growing apace. I pick a colander full of all three each day to use in salads. Mixed with the small round lime green butter lettuce that thrives in these cooler days, we munch away like rabbits, but I fear the colder days of July will bring frost-filled mornings and spell the end of salad crops.

I love lentils, so I make my favourite lentil salad with chopped coriander for lunch. I first tasted this dish on a balmy night in a smart restaurant on the French Riviera. That was years ago when, in my previous life as a fashion buyer, I flew around the world every few months. Now, as I spoon the salad onto a plate, I yearn for the long summer evenings once again.

Throughout my life I have been a frequent visitor to France. During my fashion years the visits were for business. More recently they have been for summer holidays. I am never disappointed, and I always return with a notebook of new recipes and food tips. I find the people so generous and sharing. On a recent visit to the mountain region of Provence, we found ourselves sharing an evening carafe of rosé with a local farmer. In France, the rural folk have an immense knowledge of the agriculture that supports their culinary culture. He invited us to visit him the next day to see how he distilled his lavender.

What an experience watching huge trucks of blooms arriving at an enormous still to disgorge their fragrant cargo. As one truck arrived, the previous one would be departing for another load. This all took place in the middle of a plateau covered in lavender fields. The scent of lavender hung in the air like a perfume. Later, as we enjoyed a delicious lunch at a small bistro, the scent still hung in the air like a swirling mist that enveloped everything and everyone. Later that evening whilst I was undressing, the scent was released again from the fabric of my clothes as they fell to the floor.

The food of France never fails to please. In the high mountain ranges and remote countryside you can always find a small bistro brimming with activity and offering a really good lunch for a few Euros. This is the most honest food you can find. Often prepared by a *grandmère,* it will be traditional fare prepared in a time-honoured way using the best of local seasonal ingredients.

I have decided to try a few late crops of salad and herbs planted in pots. My plan is to grow the plants that I use almost every day. I spend the morning dragging clay pots of all shapes to a sheltered patio outside my office. I've selected this spot because it's an area that catches the winter sun. I clean the pots out with a hose and wheel in a load of compost. The gravel patio is hard going for the wheelbarrow and by the time I've filled the pots with compost the morning has evaporated. It is afternoon before I actually get around to planting anything. In go some chervil, parsley and a selection of salad leaves. I cross my fingers.

This is a lot of work for a bowl of fresh salad! But I decide the pots should stay there as they'll be a picture in summer when they overflow with tomatoes and basil. If I move a table and chairs there it will make a lovely little enclave. Why is it that one job always leads to another? June is a good month to

take cuttings of trees. We take a couple of hundred Chinese poplar cuttings to propagate as wind breaks, and I decide to try my luck with almonds. I snip a few stems off before the blossom begins to set, and place them in pots. On my way back to the house later that afternoon a smile fills my face as I notice tiny shoots poking up from the trial spelt patch I planted last month.

At the bottom of a basket in the kitchen I find some rather sad-looking potatoes. Soft, forgotten and forlorn, I toss them into a flat box and take them to the store. In two to three weeks they will have sprouts and I can plant them out. Hopefully, by October, when the weather warms, I can harvest tiny new potatoes. Nothing tastes better than a freshly dug potato. If the weather is bad when I plant, I will place them in a trench on a layer of newspaper to protect them slightly from the cold, and the paper will eventually compost itself into the soil.

The potato began life growing in the poor-quality soils of the South American Andes, so they are very happy to grow in sandy soils. As with the tomato, the Spanish were the first to stumble upon this vegetable that has become a staple diet of people throughout the world. I have a friend in London who always grew some potatoes in a compost-filled black bin bag with holes punched in it. He loved the taste of a freshly dug potato and since a tiny balcony was the only space available, he found a way to satisfy the yearnings of his taste buds. He now has a cottage in the country and grows several varieties. Oh how I wish we had access to a larger assortment of potato varieties here in South Africa. It would also be so nice if people would grow more vegetables, it is not hard, and the taste of freshly picked produce is better than anything you can buy.

It's racking time in the cellar again, and I spend a day swishing out barrels in the drizzling rain outside. The wine in the older barrels has not developed as it has in the new ones, so now I pump it all into a tank and back into the various barrels. This is because each barrel of wine acquires its own distinct flavour. As the barrel itself ages, it tends to give less of a woody taste to the wine. A barrel can be used four or five times before it is too old to offer anything to the wine in the way of flavour. This suits Nic, my eldest son, as he has a little home industry of making chopping boards and platters from barrels that are past their prime.

The idea of racking at this stage though is to even things out by transfering the wine from barrel to tank, letting it all mingle and then pumping it back into barrels. It's back-breaking, time-consuming work. Luckily, it will be October before we have to repeat this process all over again.

That night a storm rips through the valley, and the lights begin to flicker while I'm in the bath. I quickly extract myself from the water, dry off and just manage to find my pyjamas before we are plunged into darkness. I find my way to the bed with the aid of a torch. Power failures are a common occurrence out here in the countryside and when we moved here we did so with foresight, installing a large gas range. Power or not, there is always food on our table and my giant copper preserving pan provides hot water in any emergency.

The longest outage we have experienced took place a few years ago. It lasted three days. The washing was stuck in the machine, the fridge slowly warmed up, and the food in the freezer became a soggy mess. We kept a large fire burning in the hearth and lived and ate by candlelight; I kidded myself that it was romantic and reminded everyone that more than half the earth's population live without the luxury of electricity. Today's power cut is short-lived and with much beeping and whining all the appliances and computers restart. Life is back to normal.

On the 21st of June, the earth is at its furthest from the warming sun – the shortest day of the year in the southern hemisphere. From this moment the days will slowly grow longer and lighter and warmer. The South African winter starts officially on the 1st of June, and ends on the last day of August, but here we get our coldest, most inclement weather in the months from July to September. I always try to celebrate the winter solstice with a party.

If the 21st of June falls on a weekend, we celebrate with a long lunch under the winter sun. It's a wonderful reason for inviting a load of friends to gather around the table and share a meal.

This year it falls during the week, so I put some champagne, made by a chum, in the fridge and plan a simple mid-week supper. Duck terrine will fit the bill as a starter and a special treat. We eat it with spiced figs, which fill the room with the heady scent of summer.

Outside, the fig trees stand bare and stubby-armed. I battled to find fig trees in local nurseries. In a desperate bid to increase our fig crop I resorted to taking cuttings and propagating my own. First I took each cutting a few centimetres from the branch ends, taking care to cut above a node. Next, I placed them into compost-filled pots, and waited. They quickly sprouted roots and soon developed their first small leaves.

Last summer, into their third year, these thirty small trees produced their first dozen or more purple figs. I look forward to more generous offerings next summer. The original "mother" fig tree stands close by. Having given us the small trees, this giant fig tree still gives us kilos of fruit year after year.

Much cheered by the champagne, candlelight and good food we fall into bed, with the knowledge that tomorrow's daytime will be just that little bit longer.

Lentil Salad

300g	puy lentils
1	red onion, finely diced
4	cloves garlic, finely diced
1	lemon
handful	coriander, roughly chopped
4tbsp	olive oil

Serves 6   Place the lentils in a pan of boiling water and simmer for 20–25 minutes. Drain and put into a serving bowl. Add the onion and garlic to the warm lentils, grate in the lemon rind and squeeze over the juice. Add the roughly chopped coriander, finally pour over the olive oil. Mix together and serve slightly warm.

Sautéed Mushrooms on Toast

knob	butter
1 tbsp	olive oil
1	sprig fresh thyme
3	cloves garlic, crushed
500g	assorted mushrooms of your choice
3 tbsp	white wine

Serves 4   Heat the butter and oil in a frying pan, add the thyme, crushed garlic and mushrooms, cook for 3–4 minutes, tossing occasionally. Turn up the heat, add the wine and cook for a further 3 minutes to burn off the alcohol. Remove the thyme. Serve with slices of toasted baguette, spoon out the mushrooms then pour over the wine-infused juices.

Duck Terrine

4	chicken breasts
2	duck breasts
6	slices streaky bacon, chopped into lardons
1 tbsp	fresh thyme, chopped
3	cloves garlic, crushed
1 tbsp	lemon zest
1	egg yolk
250ml	stale bread crumbs

Serves 6

Preheat the oven to 180°C or gas mark 4. Mince the chicken finely. Remove the skin from the duck and chop up the breast meat. Place all the ingredients into a bowl and mix together. It's easiest if you use your hands. Line a loaf tin with baking paper, leaving enough paper to fold over the top, or a terrine dish if you have one. Press the duck breast mixture into the tin. Place in a larger baking tray filled with hot water. The water must reach halfway up the terrine dish. Bake for 50–55 minutes. Remove from the oven and place a heavy object onto the terrine to compact it. Leave this on top until the terrine is cold. Serve with crusty bread.

Spiced Fig Preserve

Makes enough for 2 large or 3 small jars

1kg	ripe figs
125ml	water
piece	cinnamon bark
tbsp	cardamom pods
1	piece whole mace
1	star anise
1	lemon, zested
250g	sugar

Cut the figs in half and place into a pan with the water at boiling point. Add all the spices and lemon zest, then turn down the heat and simmer for 10 minutes. Add the sugar and bring back to the boil. Cook over a rapid heat for 10 minutes. Pour into sterilised jars, leaving the spices in the mixture.

July

The olives are still in their brine bath and should be nearing maturity. I'm keen to taste them and I wander into my storeroom where the huge buckets have been languishing for the last few months. I pop a large black olive into my mouth. It's salty and firm. So, I take a bucket down to the kitchen and heave the contents into the sink. Out spills the brine, herbs and all the olives that are now shiny and bright. They get a good rinse before I pack them into clean glass jars together with sprigs of thyme, fennel seeds and a piece of lemon peel. I fill them with new liquid and top the contents with a splash of olive oil to seal them. The jars are stacked in the cupboard, where I'll leave them another four weeks for the flavours to infuse. A few olives have gone soft, so I pit and chop them with a mix of capers, chilli, garlic, lemon and a little olive oil. I spread this dense black tapenade onto toast and know that this is the taste of summer to come.

The reality of my winter life is that I've not seen the mountains for days. They seem permanently lost in a shroud of thick grey fog. The air is heavy and bone-achingly damp. The grass is sodden as dew drips from disconsolate trees. Everything is damp. Even the landscape is veiled in mist. The days are dark and it's difficult to be sure that the sun is even bothering to rise on these dull mornings. By three o'clock in the afternoon I light a fire knowing that the crackle and pop of dry wood and pine cones will always bring warmth and cheer into the house.

The last few satsumas hang on the trees. So soggy, they cannot be picked. The grass is covered in a blanket of brown oak leaves. We rake them up every day and dispatch them to the compost. It's a continual cycle of work. The next morning a new blanket is spread out on the lawn.

The last few lettuces have turned limp and brown, so I pull them out and fling them on the compost too. This is an unappetising time of year. The only vegetable that seems to enjoy such conditions is the Swiss chard. My boots sink into the mud as I pick large bunches of the deep green leaves for supper. I will bake them in a béchamel sauce topped with grated cheese. I eat only this for supper, and use a bread crust to wipe the residue sauce from the pan.

Life carries on indoors. I busy myself sorting seed packets and tidying kitchen cupboards. The copper pots get a good polish. The cupboard shelves are emptied, cleaned and repacked. This is a good time to try out recipes and I spend most of my time cooking. The biscuit jar overflows. Then the phone rings and the kitchen fills with the acrid smell of burnt biscuits as I chat to a client. In the evenings I sit beside a blazing fire typing out recipes on the laptop.

For me, winter is the season of reading. Throughout the long dark evenings my nose is planted in a book. I get depressed with Zola confused with Beckett, seduced by Yeats and engrossed re-reading John Fowls. And finally I return to the original food goddess, Elizabeth David.

The farmhouse kitchen becomes the hub of my home during winter. Warmed by the huge stove, everyone gravitates towards this cheerful place. Meals are shared around the kitchen table as a fire flickers in the adjoining living room. The table is the epicentre of our existence, laden with not only food but flowers, books, letters, papers and my docking station. I love to cook to music. The dish dictates the genre of music. It could be Mozart or the Black Eyed Peas. But, when I walk out, my iPod is replaced, and the volume booms out the sounds of the favourite rap artist of the day. At mealtimes, plates, glasses and candles change the tablescape from books, laptops and papers. This kitchen table, which has served us for 15 years, is used to this Jekyll and Hyde existence.

In the cellar, the temperature is so cold that the cooling system can now be turned off. Barrels sit in quiet rows, stoically holding their treasure. The subtle flavour of oak is infiltrating the maturing wine, and apart from topping them up every couple of weeks and checking for unnatural odours, I have little reason to disturb them.

Due to the damp conditions, the pigs remain tucked up in the stable, which they rarely leave in winter. We cover the floor with old newspapers, and they rearrange them with their feet to bury themselves deep inside their own little piggy nests. When I go down to the stable with kitchen scraps, they run out, gobble their food and dash back to their warm burrows.

A strong north wind blows the fog away and brings a torrent of rain. It rains and rains; day after day, beating down, driven at an angle by the wind. It seeps under the kitchen door and the bathroom floor is permanently covered with discarded wet clothes. Jeans are changed two to three times a day. I check on the animals several times a day and get soaked through on every trip outside. I am amazed that the horses prefer to carry on grazing on the wet grass, rather than taking shelter in the stables available for them.

Young trees strain at the angle blown by the strong winds. From the warmth of the house I hear the continual clunk of small branches falling from the overhanging oaks. And I say a silent prayer that no large branches are torn from the trees. Boots hang upside down on the boot rack dripping water onto the floor. All growth outside seems to have come to a standstill, which frustrates me, as much of my life revolves around the food on the farm. The land is too waterlogged to plant, and it's just too cold for anything much to flourish, so there's not much to do but sit it out and wait for the season to change.

On a calmer day I trudge outside in search of something to fill my empty vases. There's not much to pick except the trusty lavender flowers and an armful of rosemary stalks. Suddenly the house smells of sunshine. What a difference. I take a bag of frozen roast tomatoes from the freezer and make a simple tomato soup, blitzing the tomatoes in the blender and adding a handful of herbs and a spoon of crème fraiche. The flavour is intense. I am grateful for the bounty that nature provides without the intervention of horrid chemicals, artificial flavourings or preservatives here on the farm. What we have is good, honest food.

The school holidays are approaching and my routine will change again. The farm is resting, and there is little that can be tackled before the busy pruning season starts. We are leaving for warmer climes and the sunshine that I miss so much. I am like a child filled with the anticipation and excitement of glorious things to come. My suitcase is so much bigger than any of the others and I am mocked for its size. My husband is the ringleader here and the boys fall in behind him in a chorus of "What're you going to put in a thing that big?" How can these modern "hunters" ever understand the exquisite delights that await the 21st century "gatherer" in the markets and food emporiums of Italy and France? None of them seem to understand the feminine wisdom of travelling with a very large, very empty suitcase. But I am smug in the sure knowledge of just how full that case will be on the return journey.

Our first destination is Italy where food is the focus for all of us. The boys lick their way through a ton of gelato while I scour the food shops. There's a culinary "find" in every food store and market, and as the shopping continues unabated my husband's anthem "as long as you carry it in your suitcase" seems to be stuck on repeat. How can one walk past truffles and good olive oil?

We visit wineries and share chunks of salami and glasses of wine at ten in the morning. My husband simply can't walk past a trattoria without going in to sample the fare. There is never a word of complaint from the boys who tuck in with gusto. We move on as our hired car takes us from Italy to my beloved France. The boys change their eating focus from gelato to the spoils of the local patisseries. I disappear into another market where I'm surrounded by stalls laden with an incredible selection of fresh fruit and vegetables. Finally, the family find me at a stall offering a staggering array of varieties of basil.

The boys hire bikes and set off to explore the countryside. We all walk a lot, eager to explore some of the medieval alleyways that twist and wind between picturesque houses with windowsills bursting with bright geraniums. It's a feast for the eyes and our sense of anticipation heightens as we turn the next corner to find yet another inviting boulangerie or patisserie.

We eat three full meals a day, all washed down with copious amounts of local wine, and I'm amazed that it is only my suitcase that gains any weight. The sun sets after nine o'clock and the sky turns an incredibly luminous indigo blue. Most nights we are in bed only after midnight and, with thoughts of fresh adventures tomorrow, we instantly succumb to a deeply contented sleep. Each morning we open the shutters to bright sunshine by eight o'clock ready for more.

My notebook is full of ideas, sketches and recipes, and I juggle with piles of French magazines that I simply have to lug home in my hand luggage. They will keep me company through the cold winter evenings at home. We are tanned, healthy and rested. The boys have experienced many new sights and sounds that will broaden their outlook on life. My husband looks re-energised. Fond farewells are said to friends and a few hours later I sip on a glass of wine to cheer my sinking heart as the plane takes off. The holiday is over. It's so good to spend quality time with the family and a trip away gives me time to think and formulate new ideas. What's more, "that" suitcase is full.

We return to a white landscape. The first winter snows have fallen and the mountain peaks are thickly dusted with snow. The house is ice cold but I still fling open all the windows and doors to refresh the stale air. The dogs won't let us out of their sight and continually nag for attention. I have new recipes from

my travels and I have found new seeds to grow. I feel motivated and am brimming with ideas. We share a makeshift picnic supper; sleep in our own beds and awake refreshed.

I take a stroll down to the almond orchard, a mug of tea cupped in my hands. This is a bothersome practice as I have a terrible habit of putting a mug down when distracted. Days later I will come across a dirty half-full mug stuck in the fork of a tree. Today I place the mug on a large stone as I inspect the trees. They are covered in a blanket of the palest pink, fragrant blossoms. They are astoundingly beautiful. It's a wonderful welcome home and I pick a few branches to place in the kitchen. The perfume is so heady, so intense and so sensual; it could never be reproduced in a bottle. Every time I enter the kitchen it fills my nostrils with the promise of spring. When they wither I pick a few more. The trees blossom for such a short time I simply have to indulge in their beauty.

Clear sunny days follow and we start to lunch outside again. It's too early to plant out summer crops, but I dig deep into the compost heap and fill seed trays. That afternoon, a few seeds are planted in anticipation of warmer weather. They are placed on a metal shelf rack in a sheltered spot.

If you stand stock still near the vineyard, all you will hear now is the "snip snip" of secateurs on wood. The pruners are busy shaping the vines for a new season. The wood is cut right back to two buds while the vine is dormant. These buds will sprout with new growth in September. The air is full of the smell of burning vine wood as I busy myself pruning the roses and the apple trees. Huge mounds of cuttings follow my progress. As the afternoon sun grows weak, so do my hands, which ache from the action of pressing the secateurs. My back complains from all the stooping and, as pruning continues, I recruit help in reaching the uppermost branches of the apple trees. Hands and feet nimbler than mine will have to tame the higher reaches of the trees.

My eldest son has a birthday soon. It falls on a Friday so I prepare a special dinner. The days of parties were left behind when he entered his teens. Tonight we will feast on fillet steak with mustard sauce, followed by indulgent chocolate choux pastry buns coated in dark melted chocolate. I busy myself in the kitchen while he chats on the phone, opens presents and enjoys being spoilt and indulged.

As the month draws to a close the sunny days are replaced by more wind and rain and I am driven inside again. The Jersey cows enjoy the luxury of our sheltered top paddocks, where they're protected from the cold. They have a habit of calving in the worst weather. Perhaps it is their primal instinct to calve when predators are least likely to be on the prowl. Whatever the reason, one of our young cows, Storm, was appropriately named.

This herd of glamorous girls with their sleek golden coats and seductive long eyelashes started off 10 years ago with just two cows and a naughty young bull called Thomas. He was named by my younger son in honour of the famous tank engine of children's literature. Thomas was tiny when he arrived. Even so, his speciality was breaking out of paddocks. I remember one stormy night when supper with a friend was interrupted by loud hammering on the kitchen door. It was one of those dark rainy nights and I opened to find a man whom I did not know informing me that a young bull was walking down the road. He promptly roared off in his Toyota chariot and I bundled on my rain jacket and set out into the night leaving my friend cradling a glass of red wine. I set off up the driveway at a jog, head down to prevent the driving rain stinging my eyes, and I ran straight into Thomas trotting back down the driveway. He blinked at me indignantly as though I was supposed to know that he likes to go for a walk, but always returns home, and trotted back to his paddock. Once there, he simply hopped over the then-insubstantial fence. The paddocks have since been made secure so that the cows stay put, and I can pass my evenings uninterrupted.

The last few days of the month are spent clearing up after a storm. Broken branches are scattered all over the farm, and the rickety old bridge that was flooded by the raging river is blocked with all sorts of debris brought down in the torrent. On the day of the storm, the weather was too bad to work in, but I walked through the driving rain to check the bridge, holding the road to the far end of the farm. Through the rain I spotted a bright yellow oilskin, then another. Two staff members, who have worked with me for years, had had the same notion and we stood together in the driving rain, mesmerised by the sheer force of the elements.

Part of the garden wall has been destroyed by the flood, taking tons of soil along with it. The garden is a sodden mess and the lawn is covered with mud. We cart away loads of debris and dig over the sodden vegetable beds. Bravely I plant a few seeds (fennel, peas and lettuce), in the hope that the worst has passed. I return to browse through the magazines lugged home from France. Winter is locked outside while I lose myself in the summer images that cover their pages. I scribble notes. By the end of the evening I resolve to put the dreary winter behind me and re-plan and plant the herb gardens outside the kitchen. It is from these four squares of soil that I gather the herbs to flavour and enhance our meals. If I revamp them now they should be flourishing by summer. Although autumn is strictly the best time to transplant herbs, I need to do this now and I will plant tender new seeds and seedlings in pots until the warmer weather is here.

Sleep takes over as I finish sketching out a rough plan. I can't wait for tomorrow.

Three Cheese Chard Gratin

1	large bunch Swiss chard
100g	butter
30g	plain flour
1 tsp	Dijon mustard
500ml	milk
50g	feta cheese, crumbled
50g	Cheddar or other hard cheese, grated
pinch	freshly grated nutmeg
pinch	black pepper
50g	Parmesan cheese

Serves 4

Preheat the oven to 200°C or gas mark 6. Wash the chard well under running water, then remove the stalks. Place the stalks in a pot of boiling water and simmer for 5 minutes. Slice up the leaves, place into the pan of drained stalks and replace the lid. Leave for 3 minutes and let the heat from the pan and the stalks wilt the leaves. Drain again, leaving them in the colander while the sauce is being made.

In a saucepan, melt the butter over a low heat, add the flour and Dijon mustard and stir into the butter. Gradually whisk in the milk, then add the crumbled feta and grated Cheddar together with the nutmeg and black pepper. I do not add salt as the feta is quite salty. Stir over a low heat until this mixture is glossy and thickens, then add the chard. Pour into a baking dish and grate Parmesan over the top. Bake for 10–15 minutes until the gratin is bubbling and the Parmesan has slightly browned.

Fillet Steak with Mustard Sauce

1.5kg	beef fillet
	dash of olive oil
4	cloves garlic, crushed
1	large onion, finely diced
100ml	white wine
1 tbsp	wholegrain mustard
50ml	double cream
	salt and pepper to taste

Serves 6

Cut the fillet into even-sized steaks, drizzle with olive oil, and put to one side.

Heat a good little splash of olive oil in a frying pan and gently cook the garlic and chopped onion for about 3 minutes until the onion is soft. Add the white wine and simmer for a further 3–5 minutes to reduce the liquid. Turn the heat down low and add the mustard and cream, season to taste with salt and pepper.

Heat up your steak pan until it is very hot, and place the steaks in it to cook. They will need 2–3 minutes on each side for medium rare. Turn them only once, then take off the heat, place them on a platter and pour the creamy mustard sauce over the top. Serve immediately.

Chocolate Choux Paste Buns

makes about 40 small buns

For the choux paste

125ml	water
125ml	milk
2 tbsp	unrefined white sugar
100g	unsalted butter, cut into cubes
150g	plain flour
4	large eggs
	piping bag and nozzle

For the filling

250ml	double cream
1 tbsp	icing sugar
	dash of vanilla essence

For the topping

150g	dark chocolate

Preheat the oven to 200°C or gas mark 6. Pour the water and milk into a saucepan, add the sugar and bring to the boil. Turn off the heat and add the butter, stirring until it has melted. Stirring quickly and continually, stir in the flour until the mixture forms a soft ball. Tip this into a bowl then add the eggs 1 at a time. Beat them in well until you have a smooth glossy dough. Using a piping bag, pipe the mixture into small round balls, measuring approximately 2cm, onto a baking tray lined with baking paper. Leave a space between each ball for expansion during baking. Bake for 20–25 minutes. The buns should be pale golden brown and feel light and airy. Place them on a tray to cool. Store overnight in a tin.

The next day, whip the cream until thick, while adding the sugar and vanilla. With the tip of a sharp knife, make a tiny incision in the side of each bun. Put the whipped cream into a piping bag, poke it though the hole and fill each bun with cream. Melt the chocolate in a double boiler and dip the top of each bun into the melted chocolate. I like the chocolate to firm up before serving, so often do this last stage before I serve dinner if I am serving them for dessert.

August

It is still dark outside and I am looking through sketches and notes for the revamp of the herb garden. With a mug of tea steaming in my hand I realise how bitterly cold it is out there when the urgent scratching of the dogs at the kitchen door moves me to let them in. It is early and the farm is just beginning to glide into the day.

As it warms up, I wander down to the herb garden, sketch in hand, to survey it from close quarters. It's quite a task, but the old herbs must be dug up. Some can be put aside and replanted, but most will end up on the compost heap. The day passes and the sun sets as I lift the last of the plants out of the soil. Finally the plot is cleared, but it takes a good few minutes to stand up straight again.

Day two and the spades come out. I drive stakes into the ground and pull string between them to mark perimeters. Help is enlisted to dig up the sections of turf to be removed. This takes forever and darkness starts to fall. I am so focussed on this task that all others are pushed aside. But I can't continue to ignore the demands of family and farm and, stiff and sore as I am, I leave for town early the following morning. I'm up before dawn, because if I get a really early start I can spend an hour at my favourite city nursery before attending two meetings. I load the car with plants, close two wine deals and arrive at school just in time to collect the boys for their weekend at home. Now the fun can really begin.

Saturday morning breakfast is complete and my plants are waiting. The herbs I dug up are piled in one corner, dampened with water to keep them alive. I place the new ones next to them and pull out my plan. By afternoon one square is complete and I stand back to admire. It is looking a little empty now but by summer it will be brimming over. By Tuesday the job is done, one week and a few aching bones later. I am thrilled with the result. Gentle rain falls to water in the new plants. New seeds are sown in trays and placed on a shelf outside my office. I cover them with a sheet of plastic and look forward to planting them out.

The iris buds are bursting and soon there'll be a splash of deep purple spreading across the garden. The peach, plum and apricot trees are covered in glorious blossoms of white and pink. What a relief. There's very little else in flower, so I pick an armful of each and place them in big jugs throughout the house. The scent of the blossom fills both bedroom and kitchen and somehow the house just feels more cheerful. A tall vase of irises adds a touch of glamour to my writing table.

Out in the garden, the roses stand forlorn and naked. They are thorny sticks with only a couple of early-budding leaves to hint at life. It seems impossible that in a matter of weeks this bare wood will be covered in vivid blooms. The vines are also naked and seem to be reaching up for the wires that once supported them. Bright green grass is showing between the rows of vines and their fine blades push up through the cold damp soil. Along the river bank, hundreds of wild white arum lilies are in flower. I pick a bunch for the kitchen.

The mountain peaks are frosted with a fresh coating of snow. It's a scene from a classic picture postcard. I hear on the news that the roads are congested with cars driving up from Cape Town to view the snow-covered countryside. I open a jar of apricot jam bottled last summer and the pleasure that this simple act brings makes me realise how much my life is driven by colour, smell and taste. The colours of summer fruit, the smell of spring blossoms and the sumptuous taste of fruit are all I need to bring a smile to my face. The countryside looks beautiful but I long for the energy of the warmer months.

A blast of cold weather sends me into the cellar to rack the red wines. This must be done on a clear day or we'll end up with cloudy wine. Rainy weather, like the full moon, draws the lees up into the wine, causing cloudiness. It is really amazing to me what a powerful effect the natural elements have on our everyday lives on the farm.

By the time I leave the cellar, having re-filled the clean barrels, I am soaked to the bone. Darkness is falling and the snow-clad mountains are eerily silhouetted against the darkening sky. A sliver of new moon hangs just above one of the peaks. It is bitterly cold. Too tired to cook, and with no one else to feed, I heat up some leftover soup for myself and fall into bed.

Sleet falls steady and silent this morning. The horses have been taken down to the "passageway" paddock that runs beside the river, just like a long passageway. I know how horses hate the wind and this stretch of grassland is totally unprotected from the cold north-westerly wind. My concern for their well-being grows as the family settle around an afternoon fire. I leave the comforting warmth of the room to chants of "silly mummy, they'll be fine" and "I'll do it later" and I pull on my tattered oilskin.

Down at the paddock I struggle to pull the gate open in the strong wind, and stagger down toward the horses. They are huddled together with their haunches facing into the gale. A gust of wind shoots by and I instinctively grab hold of a post. It's howling. I forego the halters and, slinging them over my shoulder, walk up to the poor shivering hulks, giving them a friendly slap on their backsides. They don't budge and so I start jogging back up the passage urging them to follow. And they do; pushed by the wind they trot behind, following me through the gate, on through the farm yard, past scary parked vehicles and up to the top paddocks. Relieved to be in a sheltered place, they start to graze. I fasten the paddock latch and return to the house. I am completely shaken and totally soaked.

My younger son is busy making farm-style hot chocolate, by bashing a foil-clad bar of 70% dark chocolate and stirring it into a pot of boiling hot frothy milk laced with sugar and cinnamon. I rarely succumb to this indulgence, but today I eagerly sip on the velvet nectar. It restores my equilibrium and I start to plan supper. Saffron chicken is what I feel like on this wintery night, golden and comforting, ladled over creamy mashed potatoes. There are five for supper so I pop two organic chickens into the pot, knowing that by adding some wine and a few carrots, I can make an effortless soup for Sunday lunch or supper.

When my husband returned from a business trip to the Middle East a few years ago, he brought back several packets of Iranian saffron. I have been addicted to its subtle earthy flavour ever since, and always buy some when I see it. Only a small pinch is needed to transform a dish from good to sublime. A firm family favourite is saffron rice pudding, but I also add it to a baked custard and even creamy mashed potato.

All night we listen to the gale-force wind whipping around the old farmhouse. Sleep is fretful. The grass is littered with a shower of petal confetti. Delicate white and pink petals have been ripped from the blossoming trees and hurled across the farm and garden.

Each flower is the beginning of a fruit that now will never develop, and my plans for the luscious fruits of summer are dashed.

From the house I can hear the roar of water in the river. Melted snow provides an excess of water to tumble off the mountains as dramatic waterfalls. Paddocks are flooded. The mulberry tree stands in a field of water, and the valley's road is closed because the bridge is immersed in a torrent of water. I navigate the flooded roads to emerge from the other end of the valley driving to our local town to buy papers. Along the way I stop in astonishment at the sight of vine poles peeping out of a sea of water. The papers report wild seas and roofs being blown off. This, after all, is the Cape of Storms and I am grateful we have weathered this one.

Time for tea, honey and lemon-infused madeleines. This crumbly cake-cum-biscuit is a special treat, and we feast like vultures. The first batch is devoured warm while the second is still in the oven. Fuelled, the boys disappear down on the farm. My husband returns to pottering around in his workshop and I tap away on my laptop, catching up on work while the house is quiet. The boys return as darkness falls. They are covered in mud, soaking wet and laughing happily about their adventure. Leftover chicken is made into a filling soup as the men throw logs onto a blazing fire. Hot fresh bread is grabbed from its tray to be eaten as it leaves the oven *en route* to the table. Later, we sit on the couch sharing a glass of wine while the boys do homework, a rare, quiet and relaxing moment for the two of us.

We have sniffly noses and I am developing a cough, so I make a thick, creamy garlic soup, which should rid us of looming ills. I chop herbs into the soup and thicken with potato. Hippocrates knew a thing or two when in 400 BC he wrote his words of wisdom, "food is thy medicine and medicine is thy food".

I walk out to inspect the new herb garden. Nothing has grown but, then again, nothing has died either. When we moved here I planted agapanthuses down the driveway. These have multiplied extensively so I must dig them up, split the bulbous roots and replant them. The result is a huge pile of plants that must be relocated to new homes. Where to put them becomes a rather big question. I pile them up in a wheelbarrow and wheel them round the garden looking for gaps. As there aren't too many of these I plant them in pots on the front patio. Next summer they will flower and look very dignified on either side of the old door.

There's a bowl sitting on the windowsill in the kitchen, containing a very special potion. I'm hoping that it's ready, and I lift the top, stirring gingerly to see if it will release the right odour. *Eureka!* It is smelling sour. For months I've been trying to make sourdough bread starter, and here I have it. This gooey

substance has been brewing for a few days, with flour, water and a couple of raisins. I left it covered for two days, giving it the occasional stir, then removed the raisins and added more flour and water and left it to brew for a few more days.

Like a bloodhound after its quarry, I've sniffed at it and waited. I've looked for the telltale bubble and sniffed again. Now is my moment of truth, when I will know whether I have the sourdough starter for making a yeast-free sourdough bread. I immediately set about making my first loaf of sourdough bread. When I remove the baked loaf from the oven a few hours later, it's flatter than one of my usual loaves. It also took longer to rise, but it has that wonderful chewy texture of a sourdough bread. The only problem is my impatience. I was so keen to see what would happen with the bread that I forgot to keep some leaven back to form the starter for my next loaf. This means starting all over again.

I have heard of people who keep their leaven cultures going for generations. They simply keep a jar of leaven in the fridge, use what they need to bake a loaf, and store the rest. From time to time this leaven is revived by the addition of more flour and water, which is mixed in at room temperature and then left to rest before being put back into the refrigerator. Oh well, I intend to learn from this mistake because the bread was truly delicious and I'm very taken with the idea of passing on something vital that lives on to future generations.

Ever since I read the ingredients on that supermarket loaf, I've been convinced that commercial bread is all about bulking agents, preservatives and accelerators, which make it rise quicker. The aim is commerce, not nourishment. It's about bringing a consumer product to the shelf as cost-effectively as possible. On the other hand, bread made at home is done for flavour and texture and love. There really is no argument, especially when you consider that it can very easily become part of a relaxing daily ritual. Consider the way the smell of baking bread gets into every corner of a home, and the argument against baking your own begins to pale. I have friends who use bread machines and, although I'm a bit of a purist, even this is much, much better than plastic bread.

I spend the last few days of the month catching up with my writing and all the admin I have neglected. It's amazing how little odd jobs and niggling paperwork seem to accumulate. One task that is nagging at the back of my mind is to find cherry trees to plant. I spend a morning calling nurseries to no avail. In desperation I call one of the pack houses we've used to pack our fruit. They solve my dilemma and the nursery that they recommend has trees. But the trees are too small and I will have to wait until next year.

My house is always either full of people or empty, except for me. I have a husband who is away a lot. I don't mind this at all, except that it takes days getting used to having him around. Then when he goes I have to get used to being alone again. I use the time to write, safe from the danger of interruptions. The house is peaceful and quiet except for the fact that one of the dogs is limping badly and she's nagging for attention. My joints ache too. The cold seems to be taking its toll and I give Lucy some devil's claw for her aching hip and book myself in for a massage.

There are a few carrots and parsnips still growing bravely in the garden. I pull them up and fetch a small butternut from the vegetable store. I chop them all, add a couple of onions and garlic, and put them to sizzle in a pot with some lardons of bacon. From the freezer I add a litre of chicken stock and some roast tomatoes. The soup bubbles away at a slow simmer all afternoon, and is graced by a handful of herbs and shavings of Parmesan. This simple soup becomes a feast shared with a couple of friends later in the evening. I serve it with freshly baked bread and a good wedge of cheese, washed down with a glass of wine and accompanied by conversation and laughter. We share a toast to the forthcoming spring and make plans for summer outings. After a lovely evening I feel completely relaxed. Maybe I won't need that massage after all.

Saffron Chicken

50g	butter
2	onions, chopped
6	cloves garlic, crushed
1 tsp	fennel seeds
1	large free-range chicken
6	tomatoes, peeled, chopped and diced
1 tbsp	fresh rosemary, chopped
pinch	saffron
500ml	white wine
	salt and pepper

Serves 4   Place the butter in a large pot on a low heat and add the onions, garlic and fennel seeds. Sauté on a low heat until the onions are softened. Add the chicken, breast down, and brown the meat, then turn over and brown the other side. Add the tomatoes, rosemary, saffron and wine. Cover the pot and let the chicken simmer for 90 minutes. Remove the chicken and cut into pieces. Add salt and pepper to the golden juices, turn up the heat for 5 minutes, then take off the heat and return the cut-up chicken pieces to the sauce. Serve with mashed potatoes or rice.

Farm-style Hot Chocolate

	makes 4 mugs
100g	dark chocolate, in its wrapper
l litre	milk
1 tbsp	sugar
pinch	grated cinnamon
dash	pure vanilla essence

Bash the slab of chocolate on a hard surface until the contents are shattered. Place all of the ingredients into a pan and heat gently, while whisking with a hand-held whisk. When the chocolate has dissolved, pour into mugs and enjoy.

Madeleines

makes 24 cakes

2	eggs
50g	unrefined white sugar
50g	icing sugar
120g	unsalted butter
1 tbsp	honey
half	half lemon, zested
100g	plain flour
1tsp	baking powder
	knob of butter for greasing your mould or tray

Preheat the oven to 180°C or gas mark 4. Beat the eggs and both the white and icing sugar, until pale and frothy.

Melt the butter, honey and lemon zest in a pan. While these are melting, sift the flour and baking powder into the frothy sugar and egg mixture, and fold in with a metal spoon. Add the melted butter mix and stir in with a spoon. Butter your mould and shake a little flour over the butter to ensure your cakes do not stick, then place a spoonful of the mix in each mould. Bake for 8 minutes until the cakes have risen and are pale golden brown.

September

"It's the first day of spring," says the voice on the seven o'clock news on the radio. You could have fooled me. The thermometer shows that it's just above freezing. A light dusting of frost covers the paddocks. When I go to open a metal gate to take a closer look, its coldness stings my hand. I pull my sleeve over my hand and open the gate. The air is crisp and my feet make a light crunching sound on the grass.

Little splashes of vivid green are appearing here and there. Overhead, the oak trees are laden with tiny leaves that look like mice ears. Each day more and more and more appear until all the bare branches are covered in bright green leaves again. A new cycle of growth is set in motion. Lettuces, shabby with wet soil, are crisp and sweet. Small carrots have grown surprisingly quickly in the cold earth. Tiny fennel bulbs are swelling and their flavour is delicate. I add some to salads. When they grow larger I will roast them in the oven with a handful of olives as a simple supper with walnut bread and some cheese.

The day warms up by mid-morning, leaving the grass glistening with moisture from the melting frost. I am able to shed my jersey and, after a welcome mug of coffee, continue my morning stroll.

The chickens are extremely chuffed with the lengthening days; they lay lots of eggs with rich yellow yolks. Tucked up in their night-shelter they appear oblivious to the cold, laying their eggs in rhythm with the longer light-filled days. I have kept chickens for years, even before settling on the farm. Our first chicken was a bantam given to me by a neighbour. My eldest son, who was only two at the time, delighted in hand-feeding this little bird every morning. Gradually we acquired more, and have had a supply of fresh eggs ever since. We now have a selection of chickens, bred and collected over the years. They have to be locked up each night, but run freely during the day. I spot one large Rhode Island Red sitting under a thorny rose bush, protected from invaders. She will patiently sit until her eggs crack open with new life.

A few geese strut around the farm, hissing and screeching as they walk. My Jack Russell holds back; he is wary of these birds. A few ducks waddle along behind. They are the survivors of a clutch of eggs hatched on the farm by our first pair of ducks. I used to love watching the tiny yellow ducklings trail after their mother to the water. The owls and the crows loved watching them too and snatched them up for breakfast. Nature can be harsh, but the cycle of life continues. I must make a plan to get more ducks, which are useful to have around as they love to feast on snails.

The grass between the vines has grown long and soaks my jeans with dew as I walk through the vineyard. The vine leaves have grown and little buds are about to break forth. This grass, which protected the vines from the full blast of our cold winter, must now be cut so that it does not rob the vines of the nutrients needed for growth. The broad beans, planted to feed the soil with nitrogen, are in flower. When the warmer days of spring arrive they'll be feeding us with their deliciously tender beans.

I wash out hundreds of seed trays and fill them with a moist mixture of compost and sand. The time to start sewing tomatoes, peppers and courgettes has finally arrived. I scatter the tiny seeds and cover them with a soft layer of compost. They are stacked on shelves outside my office. I fill lengths of plastic piping cut in half, with holes drilled in the base, with some of the same compost and sand mix. I then sow lettuce and radicchio into these pipes and wait for the seedlings to germinate. When the seedlings are ready I can slide the contents into a trough dug into the ground. The young radicchio will be made into salads mixed with slices of sweet apple and nuts, contrasting with their slightly bitter leaves.

Working next to me two staff members are transplanting hundreds of tiny lavender plants into small pots. From there they will go into the ground. The compost heap is emptying as we spade wheelbarrow-loads of rich compost from it.

The herbs in my newly renovated herb garden are starting to grow. The weeds are growing too, and I am on my knees weeding this area for the first time. It is a nightmare. The grass that was dug out in autumn has grown back, even though I thought we'd managed to get out all the roots.

I cannot transplant the new herbs from the seed trays until I have cleared the weeds, otherwise there'll be too much competition for the young, delicate plants. Who decided the difference between weeds and plants anyway? I know I am going to have to take extra care of this little plot until it is established and strong. The sweet peas are falling from their supports, so I carefully wind the delicate tendrils around the string. There is so much nurturing to be done at this time of the year.

The horses have their halters on as we go from horse to horse with tubes of paste to squirt in their mouths. Time for deworming. The older chaps know the drill and accept the paste. The two youngsters rebel, moving their heads and pulling away just as the paste is ready to be squirted. One filly rears up and the halter snaps. She bolts around the paddock with a defiant gleam in her eye. When she stops her antics I walk up to her, give her a pat and squirt. Before she knows it she's licking down the last of the paste and I give her another pat to say thanks. Worms can cause such pain for a horse and can ultimately kill them, so we deworm twice a year as a preventative. The paddocks should be clean beforehand and all the horses be done at the same time.

As I walk back to the house our cattle hand runs to catch up with me. He is very agitated as he leads me down to the river paddock where one of the cows has just given birth to her first calf. It is stillborn. The new mother stands close by the lifeless little body. Her head is held down as she lets out the softest murmur of a moo. Her big bovine heart is breaking and my heart bleeds for her. This is the sad side of farming. The side I am not good at. The calf is buried and we keep a close eye on the mother for a couple of days. Her milk comes through as normal, but she has no baby to give it to. As she is milked in the following months I feel for her loss.

Another school holiday falls at month-end. The farm sings with the energy and noise of young people. My husband, Mark, is home for a while and there is a continual bustle of busy activity. Friends visit. The pool cover comes off and someone dives into the icy water. I am baking carrot muffins to rescue swimmers from the growling stomachs of youthful hunger. I hear shrieks of laughter followed by a mad dash to the bathroom for a hot shower. I make a mental note to stay away from the pool for another few weeks.

The guys bring baskets of wood down to the house as evening falls and the chill returns. Leisurely evenings are spent around the fire. It's a joy to see the children so completely relaxed. No tests, no homework. No worries about early mornings for now.

Down at the river's edge, the water flows faster. Our dogs dive unperturbed into the freezing water, gleefully chasing some wild ducks that make a hasty departure, while a graceful blue crane stands silently observing us. Luckily the dogs haven't noticed it. The weeping willows dip their delicate long tendrils into the water, and look as if they are drinking. These trees thrive on the marshy river bank. Every year, hundreds of arum lilies pop up through that growth beneath the willows. I don't know whether they were planted by someone years ago but these hardy white flowers always brighten up the river banks during the winter months. I love them for the constant supply of cut flowers for the house during this drab period.

The dogs love to play in the river and their excitement is palpable when they spot us heading in this direction. Of course, we can't head down there without a basket of snacks, and they know there's always a treat in there for them too. Many picnics and sundowners have been enjoyed on these banks. The children have swum, fished and rafted along these waters, now rushing to meet the Breede River on its journey to the sea. In the winter months I can hear the roar of the river from my kitchen, when it is fuelled by heavy rains and melting snows. By January the river will still be flowing, but as a small trickle that winds a path through the large boulders that line the river bank. Each year the river's level tends to be diminished by our neighbouring farmers irrigating their acres of vines and by us watering our paddocks. Right now, I am soaked by a spray of water from shaking dogs. Why do dogs do this? It seems like a bad joke on humans that they just have to shake themselves next to people.

I cut back to the house through the almond orchard and I stop to enjoy the magnificent view from this point. I notice tiny green almonds hanging on the trees. My husband and sons have disappeared down the farm on the tractor and I take my time to pass the grazing sheep, all still wrapped in their warm winter coats. They must be shorn soon. The younger lambs born last May are getting big; they dart away at the sound of my approach. I can hear them munching on the sweet new grass. This healthy diet ensures tender meat, with very little fat. This weekend I shall braise a shoulder of lamb with garlic, herbs and a few baby carrots. The pot will simmer on the stove for hours until the meat is soft enough to cut with a spoon.

The days get warmer, as do the evenings. I don't light a fire as often anymore, but I still keep a basket of wood handy should the weather turn chilly again. We never have a shortage of wood, there are plenty of cuttings from the vines and there is always a tree that's been knocked over by wind or simply overtaken by time. The firewood collected on the farm is stored in beautiful old woven wicker harvest baskets. We found them in

one of the ramshackle barns when we came here, covered in cobwebs and looking sorry for themselves. A quick dust and a light scrub made them perfectly serviceable again without destroying their aged appearance. When they aren't providing handy storage for wood, I use them for their intended purpose as harvest baskets in the vegetable garden or as a lug for weeds when I'm clearing out beds, as they are amazingly light to carry around. The modern plastic baskets we use for the grape harvets may be more practical for that purpose, but they lack the romance and heritage of wicker baskets.

In anticipation of summer I spend a day in the kitchen playing with ice cream. Our quick trip to Italy where the gelatos were magically smooth and creamy has inspired me. I remember the day when the friend we were staying with took us by boat to a little village on the waterside. It was here that we found a wonderful gelato factory making tub upon tub of ice cream that was not too rich. It simply melted on the tongue in extraordinary ranges and combinations of flavours.

I am inspired to experiment with different mixes, with eggs and without. I finally decide that the mix with a couple of yolks is best. Not too rich. I start to invent flavours. I add lavender to my honey ice cream, then ground almonds, then hazelnuts. And so it goes until I finally run out of milk. There's a huge mess to clean up afterwards, but it was worth every joyful moment.

After an early evening walk around the vegetable garden, where I'm distracted by invasive weeds, I realise I am late in organising supper. I gather together salad ingredients, toss some steaks onto the grill, deglaze with a splash of white wine, and add a knob of butter and a few chopped capers for flavour. This all takes 15 minutes. And then we can try out some ice creams for dessert.

When we first moved here I brought along lots of rose cuttings. In the intervening years I have made more and planted out thousands of roses on the farm. I have lined the road in front of the house with Icebergs. Each was planted as a small cutting. These now tower as high as the whitewashed wall behind them. I planted *rosa multiflora* with its little pink roses along the border of the farm. This was started from a dozen cuttings.

Years ago, before we came here, I was walking past a wine farm pushing my son in a pram. A tractor was ripping up a border of these tiny roses. I asked why and was told that roses don't make wine, and that they needed the space for more vines. Cheekily, I asked for a few of the stems from the tangled mass lying on the road. Back at home, I coaxed them to life, and brought them to this farm years later. They now climb up the fence and I am weaving the new growth into the fence and tearing at my skin in the process.

I carry on in the knowledge that by November the wisteria is about to flower. I love the scent, but when I planted these ones out I had to do so in secret because my husband hates this particular plant. He says they are killers. I make sure they are cut back often so he has no excuse to keep harbouring a hatred of them. Under the wisteria is tiny Love-in-a-mist and salvias that will soon flower. The Love-in-a-mist is so trouble-free it self-seeds and pops up each spring and autumn without fail. Effortless joy.

As I take a stroll with the dogs, they bound around my feet. We cross the river and walk down the farm to the far end of the vineyard where I sight a magnificent eland antelope, grazing silently. I stop in my tracks, and call the dogs to my side, the eland is a beautiful sight, and I rest on my haunches to admire him. I have seen game on the farm before and am sure that in the summer months they come to help themselves to my grapes. But I have never been so close. The dogs grow restless at my side, and the eland is alerted. He looks up, glances my way, then departs, quickly and silently. The dogs set chase, but return when I call urgently. The eland had such a dignified aura about him, I don't want him disturbed. I am happy to know that such beautiful creatures feel free to share this bounteous land with us.

I travel to Cape Town for a wine tasting at a client's shop. His customers have a definite spring in their step, and talk of summer days and what to eat and drink. Everyone seems to have woken up from a long hibernation. Several restaurants have re-opened after their winter sabbatical and I meet a girlfriend for lunch. There's an air of positive energy around the city and I enjoy the vibe of the place. On my drive home later that evening I feel upbeat and energised by the city rush. As I get nearer to home, the imposing outline of the mountains against a ruby red sky makes me realise how lucky I am to live in such a beautiful country. The dogs greet me with their usual relish and we take our habitual evening walk together.

The last days of winter have passed. I no longer light a fire, and as I open my eyes each morning, I am greeted by clear sunlight shining through the chinks in the shutters. Under my bare feet, the tiles feel warmer, and there's no longer the need to swaddle ourselves in layers to keep out the dawn chill. When I stroll through to the kitchen to get some coffee going, it is simply flooded with sunlight. At last I can open the door to a sharp morning with its herb-scented breezes, instead of blustery chill. Even the dogs are happy to root around outside instead of jostling for a position near the stove.

We will no doubt get days of rain in October, but I'm not going to fret over that too much as it will be needed. The days are stretching out, becoming longer again, and our lives are slowly moving from the inside to the outside. It won't be long before the farm is productive again. Soon I will be harvesting healthy offerings for all of us to enjoy around the table.

Oven-roasted Fennel with Olives

3	large or 6 small fennel bulbs
1 tsp	fennel seeds
50ml	olive oil
2	cloves garlic, crushed
50ml	white wine
handful	olives, pitted
2	rounds feta cheese, broken into small pieces
	sea salt

Serves 6

Preheat the oven to 180°C or gas mark 4. Cut the fennel bulbs in half lengthways if small, or quarters if large. Trim off the green tops, chop the fronds and keep some to one side.

Place the fennel seeds in a large ovenproof frying pan and toast on a low heat for a minute to bring out the flavour. Pour in the olive oil then add the garlic and the fennel. Pour over the wine. Place the pan in the oven for 20 minutes. Remove from the oven and add the olives, feta and a handful of the green fennel. Cook for a further 10 minutes. The fennel should feel soft when you prick it with a knife. Sprinkle over a little sea salt and serve.

Walnut Bread

	makes 1 large loaf
10g	dried yeast
270ml	warm water
300g	white cake flour
200g	brown bread flour
1 tsp	salt
1 tbsp	honey
50ml	walnut oil or sunflower oil
50g	walnuts
50g	dried figs or dates
1	egg yolk for glazing

Preheat the oven to 190°C or gas mark 5. Dissolve the yeast into the warm water.

Place both the flours together with the salt and sugar in a large bowl, make a well in the centre and pour in the water and yeast. Mix, then add the honey and the oil. Mix well. Add the nuts and fruit, and knead for 5 minutes. Cover the dough and leave in a warm place to rise for 45 minutes. The dough should have doubled in size. Knock back and place onto baking paper on a baking tray, forming an oval shape. Brush the top of the loaf with beaten egg yolk.

Using a sharp knife, make a couple of slashes across the top of the loaf and leave it to rise again for about 20 minutes. Bake for 35-40 minutes. The loaf should be golden brown and sound hollow when you give it a knock.

Braised Shoulder of Lamb with Vegetables

2	large onions, finely chopped
6	garlic cloves, finely chopped
50ml	olive oil
2kg	shoulder of lamb, or leg
500ml	water
500ml	white wine
4	large tomatoes, peeled and chopped
half	chilli, seeded and finely chopped
2	bay leaves
	sprig of rosemary
	salt and pepper to taste
1 tsp	sugar
300g	baby potatoes
300g	baby carrots, scrubbed and topped

Serves 6

Place the onion and garlic with the oil in a stewing pan and cook on a medium heat until soft; this will take 3–5 minutes. Remove from the pan and set aside.

Place the lamb in the same large pot and seal the meat on a medium heat; this will take about 3 minutes each side. Remove the meat from the pan and put aside. Pour in the water and wine, scraping the bottom of the pot with a wooden spoon to deglaze. Return the lamb, onion and garlic to the pot; add the tomatoes, chilli, bay leaves, rosemary, a pinch of salt and pepper and sugar. Sugar really brings out the flavour and sweetness of lamb when cooked for a long time. Braise on a simmering heat for 2 hours. Leave to cool.

When cool, scoop out the solid particles of fat that will have risen to the top of the stew. Return to the heat, adding the baby potatoes and baby carrots. Simmer for 40 minutes, with the lid removed. Check the seasoning. Serve with rice and green vegetables.

October

I wake to find the oaks wrapped in cloaks of bright lime green. The colour is blinding and I'm sure that if I were to paint a picture of this using the same lime green, it would look completely surreal.

Growth takes place so quickly, you can just about see plants stretching themselves into the new season. Yesterday's buds are today's blooms. Last week's seeds are now tiny plants. In this process Mother Nature sets the pace and it is hard to keep up. The trees blossom in an explosion of colour, closely followed by an invasion of scent, so dense that it's like suddenly finding yourself inside a Parisian *parfumerie*. The pollen is so thick that it seems to hang in the air like a haze. The nose twitches and then itches in a prelude to a fit

of sneezing. How can Mother Nature be so darn cruel, extracting the price of hay fever in exchange for the beauty of spring.

The buds are bursting open on the vines, and the new shoots are snaking along the wires like a five o'clock express train. Every day new shoots emerge as you watch. Baby lettuce seedlings pop out of the earth, young tomato plants in their trays reach out to the sun. Where does all this energy come from? Spring has finally arrived, and the plant world is dancing with joy.

I grew up in the south of England where my parents owned the quintessential cottage garden. One of the things that I loved most in the garden was an old paving stone engraved with a poem that read "A ray of sun for pardon, the song of a bird for mirth. You are closer to God in a garden, than anywhere else on earth." Every time I walked past the old garden gate I'd stop and read this little rhyme. I think of it now as I look around my own garden. A garden is a special place, and this one always fills me with a feeling of peace and well-being. At this time of year, after the hibernation of winter, I would love to be out in my garden all day.

This part of the Western Cape comes to life in spring. Wild flowers of every hue bloom in the most unexpected places. It's so beautiful, it literally takes your breath away. Better still, the mornings are getting lighter and the animals respond happily to the lengthening days. Our flock of golden Buff Orpington chickens are laying eggs every day. Out in the meadows the sweet spring grasses are encouraging the cows to give us more milk. As the sun peeps over the mountains the "girls" are to be found at the gate lowing impatiently for the relief of empty udders and fluttering their long Jersey eyelashes at the bull who protectively accompanies them on their morning walk.

We have only a few cows, so our dairyman milks them all by hand. The task is done to the accompaniment of his magnificent baritone singing voice. On our farm the dawn chorus is the trill of the resident Cape robin chat that is the proverbial early bird, and the melodic tones of traditional African song that ring out across the farm to the accompaniment of

clanking metal buckets. I am sure the cows love it as much as I do. Their milk is as rich as single cream, which we bottle straight from the cow to use in the house. When a bottle is left in the fridge overnight, it gathers a crown of rich cream a few centimetres thick. Our children have grown up drinking this nectar and find it difficult to drink shop-bought milk. In South Africa it is illegal to sell unpasteurised milk for human consumption, which is sad, as once milk is pasteurised it loses many of its natural nutrients.

When we have extra milk I often make cheeses and ice creams. My favourite ice cream is one that I plan to make today from a honey and vanilla-infused custard. I will be using this morning's fresh milk. The boys will once again have hordes of friends over this weekend and honey ice cream is always a popular item on the menu.

The beekeeper in our family is my husband Mark, who keeps a few hives down in the almond orchard. I am fascinated by bees but have stayed well away from the hives since being stung by an angry buzz of wild bees when we first moved to the farm. I think I got between them and their hive. Whatever their reason, the painful memories of the side effects of that attack keep me from venturing near the hives again. However, I can look on with amusement as Mark dons swathes of clothes, which he tops with a ridiculous hat before sauntering off to the orchard. The following day he will again disappear into the bowels of the farm shed to reappear hours later laden with jars of the most wonderful honey that never fails to please. I think that the quality of his honey is due to the immense care that Mark invests in the bees and to the wonderful food supply they have on the farm.

I welcome the early morning light because I hate the dark. However, as we dig the soil for our next crops of potatoes and beans, we find it is still dank and cold from the long wet winter. The peas seem to revel in this damp cold earth and I notice baby pods forming in a rush of new life encouraged by the spring sun. I find myself picking broad beans every day. There has been such a glut of them that we have been able to feast on these early gems from the garden. I notice, though, that they are starting to become big and woody, so I harvest the rest of the crop for next year's seed. Now, too, I get down to making wigwam tripods from long bamboo canes, lashing them together with string, in preparation for the glamorous borlotti and runner beans.

Each day I plant seeds in trays then stack them on shelves in the sun. Basil, peppers, tomatoes and aubergines sit in neat rows. Already in the ground are lettuce, beans and courgettes; there seems to be no end to the planting.

From the direction of the cutting garden, I catch the strongly perfumed scent of sweet peas rising from the cutting border to waft over a haze of multicoloured Love-in-a-mist. I love that scent, even though it makes my nose twitch helplessly. The first flush roses are about to open and to fill the garden and our vases with a gorgeous display of nature's bounty. Life outdoors is perfect and makes the inside look decidedly gloomy. Time for a spring clean. Down come the curtains, out comes the polish. Tomorrow I will attack the rugs. My energy levels are receiving a spring boost.

At the village cooperative, I overhear farmers discussing the treatments they plan to spray on their vines. This fills me with trepidation. Why treat healthy plants? It's rather like putting an athlete on steroids. He may run faster, but for how long? I pay for my bundle of string and retreat to my green oasis of plants ... and weeds.

This morning I pick the first peas. They are so sweet, I just can't help gobbling them down as I pick. Straight from the pod they are like sweets and it's a wonder that enough make it to the supper table. The first sprigs of tarragon push their heads up out of their winter retreat. This wonderful herb makes me think of hollandaise sauce. But tonight I think I will make tarragon chicken to celebrate this herb's arrival.

Life on the farm is one of constant renewal and retribution. In this cycle, when livestock reaches maturity it is time to make way for the next generation. It is something that I have had to get used to as I watch our fluffy buffy hens scratching happily around the farmyard and garden. The pullets, all young and sleek, are just starting to lay their small eggs. These young birds now produce one gorgeous egg every day while the older girls lay one, miss a day, miss another and lay again. I love to watch them sorting out the pecking orders as the new girls start to assert themselves. It's a bit of a hard truth to face, but in my heart I know that the old girls and surplus of cockerels, who are about to crow, will be really good cooked slowly in a casserole pot with tarragon.

One chicken is poaching in the pot and as its succulent flesh is falling off the bone, I peel it off with a fork and toss it into the waiting creamy tarragon sauce. With the plump peas picked from the garden this morning, there will be a feast for the two of us and a couple of friends. I think I'll bake a crusty baguette to help wipe our plates clean, and we'll wash it down with ice-cold chardonnay. This will be our first taste of spring.

The citrus orchards are in blossom and the scent is so intoxicating that it beckons me over the river towards the orchard where the trees are celebrating spring. They are all dressed up in their wedding dresses of white blossoms. It's amazing how each tiny blossom can emit such a powerful perfume. Neroli oil, used in the perfume industry, is distilled from orange blossoms. You must need millions of these tiny flowers to make a small amount of oil. And you would have sacrificed all the fruit in the process.

In the vegetable garden, the first tiny tomato plants are transferred to the soil. Six different varieties are planted, a small red cherry, a pear-shaped red and yellow tomato, the Roma plum tomato, which is good for preserving, and two large red beefsteak tomatoes, which will be glorious sliced with mozzarella, herbs and lashings of olive oil. Three more seed varieties are waiting in trays. I fetch black wattle stakes. The wattle tree is an alien invader that we cut down. Their thin wooden trunks are useful as supports in the garden and I drive these hard into the soil with the back of my spade. It is satisfying work. Carefully placing a small plant next to a stake, I give each one a sprinkle of water as I go. With sixty tomatoes done, I will put out another row next month, mixing the varieties to ensure a constant supply.

Michael, my trusty garden assistant, works alongside me, painstakingly weeding rows of strawberries. The four rows will take him a couple of days and once this is done I do a little companion planting. Onions were planted as seeds in my small propagating nursery. Now they will be thinned and transplanted to the rows between the strawberries. Onions and strawberries are such good bedfellows that they have a symbiotic relationship. Onions grow plump and sweet beside the strawberry plants and their strong scent protects the berries from marauding insects. It's a match made in heaven, as is the glass of wine that beckons me indoors after the day's work.

My arms and back are aching now, so the rest of the day is spent researching and reading and longing for a herb-infused bath.

The herb gardens are full of fresh new green growth. Many of the plants that have died back in winter, such as the chives, now push their way up through the soil. Just imagine the strength of a small shoot cutting a path through the soil. The sage plants no longer resemble sticks as furry little leaves develop. Even the parsley is growing before my eyes. I pick sprigs of everything and smell them. The aroma is indescribable.

As I walk across the farmyard, a swallow swoops past my head. These long-distance frequent-fliers have just returned to their southern homes. I recently read a wonderful book following their amazing route from Wales to South Africa. Each year I marvel at their ability to fly across continents, and am so happy to welcome them home. After we built the new barn in the farmyard, one made a nest behind one of the huge beams. For eight consecutive years since then, a swallow has returned. JJ my Jack Russell cuts a comical figure as he chases their shadows across the farmyard.

The north wind is picking up. Why is it that every year, after I've planted out tender young seedlings, a storm blows through to taunt them? The next day is even colder. The exuberance of spring is dampened by the sudden return of winter. Heavy rain starts to fall. Finding an old packing case I staple heavy plastic to the open sides and drag it to the sheltered patio outside my office. I am hopeful that this makeshift "greenhouse" will provide the trays of seedlings with some protection. The young tomatoes are being blown about and I fear for the delicate white blossoms covering the satsuma trees. I run down to the stables to usher the pigs under cover. Pigs can't handle a sudden change in temperature and we lost two young sows during last year's cold front. With the pigs safely inside, I drag a basket of wood into the house and light a fire. Finally, I peel off the clothes that are stuck to my wet body like a second skin. Within thirty minutes I have bathed, changed into evening clothes and make-up, and drive carefully through the driving spring rain to a friend's birthday dinner. I do hope they were not planning to have it outside.

I awake to silence. The rain has stopped drumming on the roof. I look out of the window to see that the clouds have cleared and the mountain is alive with waterfalls. Everything is sodden and the tops of my young tomatoes are gasping in the middle of newly formed puddles, battered and bent but still alive. Spring has returned and everything looks even greener.

Earlier in the year, I moved some pots to my new patio and filled them with salad crops. I've produced many a salad from these pots, which now look sad and bare. The salad crops have long since been eaten and all that remains is a motley fringe of parsley. I spend the day happily sprucing these pots up for summer. Small red and yellow tomato plants replace the lettuce and I add chervil along with purple basil. The pots will be a blaze of colour under the intense summer sun. The green garden table and chairs are cleaned, and we sit there in the early evenings with a glass of wine.

The house is full of children again. My boys have brought a bevy of friends home for the weekend. At supper I worry at how few of them eat the salad I serve with the meal. I try to tempt them and out comes a list of foods that one or another doesn't like to eat. I can understand a child not liking Brussels sprouts, but this was an assortment of fruit and vegetables. How can their small bodies grow up strong and healthy?

My eldest son regales me with stories of schoolmates who devour chips and fizzy drinks at first break, and my shock turns to horror. It strikes me that a large percentage of today's youngsters are hurtling towards a health crisis. Many will never lead normal, healthy adult lives if eating habits like these continue unchecked. I am filled with sadness that some of these children will never know the fun of sitting around a large table laden with tasty healthful foods.

In his seminal 1825 book, *A Physiology of Taste*, Brillat-Savarin wrote these prophetic words: "Tell me what you eat, and I will tell you what you are". What will become of us if we fill our bodies with rubbish? I believe that children who understand food will grow up stronger and more capable adults. Never one to give up, I make a raw broccoli salad with fruit and nuts, chopped bacon and a honey mayonnaise dressing for lunch the next day. They try it and eat the lot. There may be hope yet.

October is the birthday of Little Dance, a special horse that lives on the farm. Ten years ago, when we first moved to the country, I got a call from someone I knew down at the SPCA. They needed a home for a horse and her newborn foal. Little Dance was the foal. Her mother, Uno, had been badly hacked with knives by a Cape Town gang. This horrendous act had been part of an initiation ceremony to prove the "worthiness" of prospective new members. Uno was in a pitiful state when she was rescued by the SPCA. Her wounds were slow to heal, she was badly lame and so thin that the birth of Little Dance came as a complete surprise to her caregivers at the SPCA.

Mother and daughter arrived on the farm to recuperate. Here, they settled into a life of love and care in a place where there is plenty to eat. Little Dance is one of my most special farmyard friends who has grown up to be loving, gentle and strong. In the intervening years she has given birth to two of her own foals. She is happily grazing in the paddock with the other horses as I walk over to give her a birthday hug.

Honey Ice Cream

makes 1 large bowl

800ml	single cream
2	egg yolks
150ml	honey
	a few drops of vanilla essence

Warm the cream in a pot, but do not let it come to the boil. In a separate bowl, beat the egg yolks until creamy, then stir in the honey and vanilla. Add this mixture to the cream and stir gently until amalgamated. Cool. When cold, place in an ice cream machine and freeze.
If you don't have an ice cream machine, place in the freezer in a container with a lid. Take it out of the freezer every couple of hours and stir well, until the mixture is frozen.

Serve with a macaroon or a shortbread biscuit on the side. It also tastes good with fresh figs.

Tarragon Chicken

1	organic or free-range chicken
500ml	white wine
500ml	water
2	cloves garlic
	For the sauce
50g	butter
30ml	plain flour
500ml	poaching liquid
100ml	cream
handful	tarragon, chopped
	salt and pepper

Serves 4   Place the chicken in a large pot with the wine, water and cloves of garlic. Bring to the boil, then simmer for 1 hour. Remove the chicken from the poaching liquid and put to one side to rest. In a separate pan, melt the butter, stir in the flour to make a smooth paste, ladle in 500ml of the poaching liquid, then add the cream. Stir over a low heat for 5 minutes until the sauce thickens slightly. Add the chopped tarragon, and season with salt and pepper. Roughly carve the chicken into portion-sized pieces. I usually leave the leg and wing bones in but remove the skin. Place the chicken on a platter and spoon over the sauce.

Serve with some boiled potatoes and a green salad.

Variation: If tarragon is not in season, I have also made this dish using parsley.

Broad Bean Purée

1kg	broad beans
2	cloves garlic
½ tsp	cumin seeds
	juice of half a lemon
2 tbsp	crème fraiche
	salt and pepper to taste

Serves 6 — Pod the beans and plunge them into boiling water. Tender young beans will take a couple of minutes, older ones 3 to 4 minutes. Drain and remove the outer layer of skin.
Put into a blender with the garlic. In a shallow frying pan, quickly toast the cumin seeds to release the flavourful oils. This should take about a minute. Add the cumin to the beans, and squeeze in the lemon juice. Blitz to make a smooth paste, then spoon into a bowl and stir in the crème fraiche. Add salt and pepper to taste.

Serve on brochette or bread as a starter. If serving as a starter, I often place a small handful of olive oil-dressed rocket alongside on the plate.

Tip: In this way you can use up older broad beans that are not so tender.

November

November is my favourite month of the year. Summer really only begins with the rose blooms, and I love the way the petals open tentatively as tight buds then unfurl into rich whorls of velvet petals that reveal the tight yellow knots of stamens within each rose's heart. The air fills with their sweet scent and I simply cannot resist picking armloads that I use to fill the house with glorious colours and perfume. There are vases of them in each room, which is a total extravagance, but I cannot resist. On some days I collect the petals early in the morning before the scent has evaporated and carry them to the kitchen to make rose water.

My grandmother always made chamomile water from the plants to use as a hair rinse. Throughout her life she also collected rain water, which she used to wash her hair, claiming that normal water was too harsh. As a child in England we would often drink elderflower water made from the beautiful little flower blossoms. My favourite flower water is lavender. I use this heady tincture in the washing machine as a substitute for fabric softener. The scent lingers in the linen cupboard and on the days when our sheets are changed it's an added pleasure to climb into a bed that releases the gentle whiff of lavender as the sheets are drawn back. I also add it to the bucket when washing the bathroom floors. It's much nicer to fill the house with the scent of flowers than the scent of horrid chemical cleaners.

During the summer, rosemary water tends to keep the flies at bay, and lemon freshens things up. I am fortunate to have a steady supply of lavender water, which is a handy by-product of distilling. You can make your own by mixing a few drops of good-quality lavender oil to a few drops of vodka and diluting with water. Rose water is also simple to make by boiling the petals with water. I use this to flavour summer desserts, but it is also a nice addition to the bath.

The largest selection of flower essences, oils and water I have ever seen was at the mountain-top village of Gourdon, close to Grasse, in France, where they distil a huge variety. I purchased verbena oil that has a heavenly scent and this year I am going to make verbena water and syrup from my own lemon verbena bushes. I think I shall spoon it over a light sponge cake.

I have been collecting rose bushes for years, the best being the old garden varieties and the English roses. The latter is a cross between the old-fashioned roses and the modern rose. The result is the voluptuous blooms and heady perfume that makes the rose a true garden queen.

There is a beautiful tale about how the red rose became a symbol of love. From the mists of time comes the story that all roses were once white. One day a nightingale was overcome by the beauty of a rose. He courted the rose by singing to the bloom, but as he pressed his body close to the flower a thorn pierced his heart and his blood stained the rose red forever.

The evenings grow lighter and we are able to spend more time outdoors. It is chilly but beautiful. The mountains are outlined with gold by the fading sun, and our evening meals taste all the better for being eaten outside.

Spring-cleaning fever takes hold of me once more. This time I focus on the exteriors. All the outside furniture is washed and polished. I plant out pots of herbs and pansies and place them on the patios. Just as I get used to this happy routine, the weather turns again and that perennial culprit, the north wind, brings in the rain and the cold, plunging us straight back into winter. My new plants shudder under the deluge and all the wine farmers live in fear of downy mildew.

This malevolent disease thrives in damp conditions, affecting first the leaves, then the berries forming on the vines. The disease first shows itself on the young vine leaves, appearing like an oily spot that takes hold and spreads so fast that it can wipe out an entire crop. Downy mildew is the curse of the vintner, putting fear into the hearts of all who live in the valley.

Many noxious sprays are currently available to delay and kill the spores of this disease, but I fear that this is not all that they kill. On our organic farm we have to be vigilant, and ensure that the canopy of the vine is kept open so that the wind can easily pass through, inhibiting the spread of the mildew.

We patrol the rows of vines, checking the new growth and taking out unwanted shoots. The temperature has plummeted, slowing the growth. Last year was cold too, with snow on the mountain, which did not melt until November. I delay plans to plant out butternuts and sweet corn, and fear yet again for my tomato and basil seedlings.

What rough weather we are having this year! The peach and plum trees are taking a battering, with much of the underdeveloped fruit lying battered on the ground, never to reach maturity. I place some trays of delicate basil seedlings on the kitchen shelf in an effort to protect them. I love this pungent herb and start many plants in trays before transplanting them to pots. At the end of the season I pound their leaves together with garlic and lemon in the pestle and mortar and store it in ice trays. At the moment I am particularly besotted with tiny-leaved Greek basil – you can pluck off the whole leaf and toss it into salads.

As the rain disappears and the sun warms up for summer, seasonal normality returns. The weeds go wild. They are everywhere, and everyone on the farm is either hoeing them out or slashing them down. I am obsessed with pulling them out of the moist soil. Why do weeds always grow faster than all the other plants? But they have their use as tons of green growth lands on the compost heaps. Left to rot down, they will provide food for the garden's next generation of vegetables and so the circle of life will continue.

I pick the first green beans of the season, then turn to the row of peas and pick some of them too. This evening we will celebrate their arrival with a simple salad combining their crisp spring flavours with a crumble of fresh goat's cheese. I pluck the spirals of pea shoots off the plant to add to the salad. Peas are a wonderful vegetable to grow as they freeze so well, without losing their texture or flavour. I always grow an excess to store for the winter months.

As the days grow warmer, I sit outside under the dappled shade of the old oak trees, tapping away on my laptop. As I pause for thought and glance out into the garden, a movement catches my eye. I notice soft black soil slowly pushing up into a small mound in the grass. The Jack Russell sitting at my side spots the activity and pounces with uncontrollable joy. The first moles have woken up after their winter hibernation and my heart sinks. From now on the grass will be peppered with soil hillocks made by these determined little creatures. I have tried everything from garlic, vibrating mole-deterrent machines and whining windmills to scare and organic potions guaranteed to deter these insistent burrowers. Nothing works. My sons have even used a hosepipe to try flooding them out. Ever-eager, my little Jack Russell is too slow to catch them. Within hours of my flattening their evidence with the lawnmower, new mounds will start to appear just to taunt me.

The next batch of summer vegetables is sown and I spend the rest of the morning tying the rapidly growing tomato plants to strong stakes. As always, my hands are stained bright yellow by the plants. I am reluctant to scrub the persistent stain off as it holds the evocative aroma of freshly picked sun-ripened tomatoes. I walk around for hours intermittently smelling my hands.

Early in the growing season, little marigold seeds are scattered on either side of the tomato plants. The pungent scent of the bright orange marigold flowers is offensive to many insects and keeps these pests at bay. I like to pick the marigolds and place them in small antique bottles around the house. I strongly believe that the scent from these little posies keeps the flies out of the house.

In the evening I sit at the little metal table outside the kitchen door, preparing the vegetables for supper. The evening air is filled with the scents of the season. Later that night I sit at the table with my laptop and a new battery-operated, energy-saving lantern. The silence of the dark enfolds me and, as I work, a squadron of moths is drawn to the light. Soon I am surrounded by dozens of little night creatures transfixed by the glow from the lantern. Time to log off and leave the nocturnal creatures to the darkness of the night.

Walking from the cellar to the tasting room I spy a golden coloured snake slithering across the farmyard. It slides along deftly in front of me, slipping over the wooden gate and up the drive. I follow cautiously. I hate snakes. But have to admit that this golden-yellow Cape cobra looks quite beautiful as it moves ahead with such grace and ease. We often see the cobras on the mountain; I wonder what has brought this one down here. The snake disappears into the undergrowth and is gone, back in the direction of the road. The warming weather has brought it out of hibernation. On hot summer days you often spot snakes in the middle of the road, which is a dangerous place to sunbathe. We have seen two types of snake on the farm, the puff adder – slow, lazy and deadly; and the Cape cobra – aggressive and lethal. For the next few days I keep making a point of looking down in front of me, expecting to see it again.

When we arrived here, I planted dozens of *tulbaghia*, the wild garlic. They have a lovely tiny lilac flower, with an aroma of garlic. They grow in clumps and have to be split up every couple of years, so the dozens of plants have now multiplied to hundreds. I often add them to my flower arrangements. These delicate little flowers are also planted amongst the agapanthuses, their big brother

look-alike. They make a beautiful border for the roses and I've interplanted them with the rose geraniums. This flower has the most gorgeous rose perfume. I love to crush its leaves between my fingers so the perfume lingers with me all day. I remember a friend baking a rose-scented sponge cake by placing the leaves at the bottom of her baking tin. It was simply delicious.

The bees love the *tulbaghia* and buzz around them all day. In the early evening I gather the wild garlic leaves and snip them up to sprinkle them over a salad of baby leaves, dressed with olive oil. The sun is setting as I'm collecting these and its rays make stripes of lime green and gold that wash over the vineyards with an extraordinary effect. It is a stunningly beautiful scene.

Dozens of small birds are fussing around the huge oak trees feeding on the sticky honeydew that the oaks are dripping onto the outdoor furniture. It is very frustrating, for as I clean it becomes sticky again. While I clean, my son flies around me making circles on his J Board. He's as free as a bird.

I love it when plants seed themselves. I have odd plants popping up everywhere like rebellious teenagers. Potatoes grow on the compost heap. Beans relocate to the herb garden and they may have to move again. Foot-free tomatoes are shooting up all over the place, and the other day I found an onion in the rose garden. Only God knows how it got there, but it's proof that they make good companion plants. I'm not a terribly orderly person so this natural planting by Mother Nature does not bother me. With the exception of the rogue bean, the plants are free to remain in their chosen homes.

Tomorrow I must load up my hatchback with boxes of wine to deliver to Cape Town. I must leave early to avoid the crazy city traffic and collect the children on my way back. The whole family will be home and I am looking forward to cooking outside this weekend. Before I leave, I take a leg of lamb out of the freezer. When it has defrosted I will carefully remove the bone and smother it in garlic and herbs.

It is late afternoon as I drive through the large wooden farm gates. The car is laden with children, dirty laundry and a stash of lovely cheeses secreted in a coolbox below the school bags. I collected these earlier from a cheesemaker friend of mine. Tired and dirty after driving around the city all day, dodging traffic, searching for parking and heaving wine boxes around, I am relieved to be home in the country. So too are my happy, hungry children. Some friends are popping in for supper later this evening.

The boys are in the pool, oblivious to the still chilly water, and filling the garden with the sound of their pranks and chatter. Meanwhile the lamb that is to be our dinner is soaking in its marinade and I walk around the vegetable garden unwinding from the stress of the day. A familiar chat is perched on top of a stake, flicking his wings to the trill of his sundown serenade. I stop to watch and notice the first young courgettes. They weren't there earlier this week. Now the golden flowers that adorned the vines have sent forth their succulent baby marrows. The vegetable problem for the evening is solved as I pluck them from the vine. There are more strawberries as well and I pick a bowl full. They are small but sweet.

I emerge from the shower refreshed and ready to welcome guests to our table. My eager young men have lit the fire, which is nearing the perfect ember stage needed for cooking. My boys love to build the small pyre that gets a good fire burning. What is it about boys and fire? Is it a primal instinct that still flickers in our souls, reminding us of the important role that the discovery of fire has played for us all?

I hear the crunch of gravel as friends arrive. The evening fills with gentle laughter as they approach the house bringing smiles and gifts of spring flowers. The last few flames from the fire dance in the fading evening light as we sit on steps or chairs chatting and drinking wine, drawn by the comforting light of the fire on this beautiful crystal evening.

Some random dishes of our home-cured olives and freshly toasted almonds sit next to glasses as we catch up on all the news.

The youngsters plan for their forthcoming holidays. I sneak into the kitchen to prepare my courgettes on the stove and set a platter of fresh bread while the butterflied leg of lamb sizzles on the fire outside. The smell of herby lamb cooking is sensational. If the weather is inclement, this dish is just as good when roasted in the oven.

We sit around the long table with its row of assorted candles flickering in the occasional breeze. The lamb, flavourful and succulent, is quickly devoured and we all mop up the juices with fresh bread.

"What's for pudding?" is the expectant call from the youngsters. I grab one of my sons to help me with dessert while the others stack dishes and load the dishwasher. Just before supper I made a bowl of fluffy yoghurt cream, which we now spoon into an assortment of old wine glass. We add strawberries to each, followed by a drizzle of rose flower syrup from the fridge. It is a feast to satisfy any sweet tooth. When I bring out the platter of cheeses collected earlier in the day, the table is full of the clutter of a good meal. We all sit well into the night chatting, laughing and exchanging stories. The men discuss politics over a bottle of red, the women exchange recipes and the youngsters share jokes.

Replete, I bid our guests farewell, blow out the candles and flop into bed, relaxed and happy with the knowledge that many long summer evenings lie ahead.

Leg of Lamb with Fresh Rosemary & Thyme

1	medium leg of lamb, bone removed (you can ask the butcher to do this)
4	cloves garlic
	zest of 1 lemon
1	large handful of rosemary, chopped
1	large handful fresh thyme, chopped

Serves 6

Cut the garlic into slithers, and insert the slices into small incisions in the meat made with a sharp knife. Zest the lemon rind and mix with the chopped herbs. Place this mixture onto a board and press onto the lamb. Do the other side as well. Leave the flavours to infuse for 30 minutes.

Place the lamb fatty side down onto a grill over a medium-hot fire and cook for 15 minutes, watching that it does not burn. If it starts to burn, lift the grill higher. Turn the meat and cook for a further 20–25 minutes. Remove the lamb and let it rest for a good 10 minutes. Remember the meat will carry on cooking a little during the resting period. Carve and serve.

Tip: The cooking times given work for a medium-sized leg that will be pink in the middle when served. If you like your lamb well done, add an extra 10 minutes to the cooking time.

Sautéed Courgettes with Mint & Capers

12	small courgettes or 6 large ones
50ml	olive oil
4	garlic cloves, crushed
	zest and juice of 1 lemon
1 tbsp	fresh mint leaves, chopped
1 tbsp	capers, chopped

Cut the courgettes into slices at an angle, and place in a frying pan with the olive oil and crushed garlic. Sauté the slices until they are turning brown. You may have to do this in 2 batches. Do not overcrowd the pan. Place the cooked courgette onto a piece of paper towel to absorb the excess oil, then arrange on a serving platter. Grate the lemon zest and squeeze the lemon juice onto the courgettes. Lastly, dress the dish with a sprinkling of chopped mint and capers. Serve at room temperature.

Rose Petal Syrup

250ml	rose petals
500ml	filtered water
2 tbsp	unrefined white sugar

Place petals and water into a pan. Bring to the boil, then take off the heat to cool. When cool, cover and place in the fridge overnight. The next day, strain the rose water into a pot, add the sugar and bring to the boil. Reduce the heat and simmer until the liquid resembles a thin syrup. Store in a bottle in the fridge.

Tip: Pick the petals early in the morning before the scent has evaporated. Pink and red petals mixed with a few white ones make the best coloured rose water.

Strawberry Yoghurt Cream with Rose Petal Syrup

50g	almonds or hazelnuts, chopped
350ml	thick cream
300ml	plain yoghurt
splash	vanilla extract
30g	icing sugar
36	strawberries, quartered
	rose petal syrup

Serves 6

First lightly toast the nuts by placing them on a tray in the oven at 180°C. This will take a few minutes, but watch them carefully so they don't burn. Put aside to cool. Mix together the cream, yoghurt, vanilla extract and sieved icing sugar and beat with a hand-held whisk until thick. The mixture should be thick enough to form a peak, but not stiff. Pour the mixture into bowls or wine glasses. Scatter over the cut strawberries, drizzle on the rose syrup and sprinkle the chopped nuts on top. Serve immediately.

December

A frenzy of anticipation sweeps through the house as the summer holidays draw near. All my paperwork is brought up to date. Customers are sent festive wishes. In the cellar, barrels are checked and topped up. I walk through the vineyards, where grapes hang contentedly, swelling to fullness on the vines. The vegetable gardens are weeded, then weeded again. The fruits of our spring labour are waiting to be consumed and enjoyed by family and friends.

The gardens look beautiful. Vegetables lie resplendent in an array of shapes and colour. The cutting border is bursting with the flowers that will grace the house and fruits ripen on the trees. What a stunning time of year this is, a chorus of joyous life.

My last chore is to make sure that the store cupboard is topped up with the essentials, and the last few Christmas presents are purchased. I hate the bustle of a last-minute rush of Christmas shopping, which always leaves me with presents that no one needs. The first week of December flies past. I catch my breath and meet some girlfriends for lunch before we all go our different ways for the holiday. We laugh, recount stories and exchange recipes for quick holiday dishes that fill hungry tummies without too much hassle for the cooks. We fret about how we will look in our swimming costumes before we part ways with hugs and kisses. En route back to the car I pass groups of tourists devouring ice creams and strolling slowly along the pavement. Summer has arrived. I stop to collect a selection of cheeses from a friend's farm before collecting the children from their last day of school. The traffic is nightmarish as hundreds of extra cars seem to have been unleashed on the roads. I am looking forward to lazy days on the farm in the company of family and friends, sharing good food and wine.

The holiday mood is infectious. The car is loaded with books, laundry and sports equipment that has accumulated during the school year. The boys are bubbling with holiday fever as they quickly unload the car, change into farm clothes and disappear in search of their first adventure.

I feel myself start to relax and switch to a slower pace. Exams are over and six weeks of school holidays lie ahead. The sense of relief in the boys is palpable. Mark's business is also winding down for the holidays. Slowly we all put experiences of the year behind us, take time to reflect, relax and cherish the quality time that we can now spend in each other's company.

We wake each morning to the sun streaming through the windows. The vegetable gardens are a riot of green growth. I drink my morning coffee outside, and take in the brew along with the peace of the fresh morning air. Then, mug of coffee in hand, I walk barefoot across the dew-covered grass to wander around the garden. The vine leaves wave gently in the faint breeze as a morning greeting. The young grapes have grown larger to form splendid hanging bunches, and I breathe a sigh of relief that there has been no damage to the crop. After the recent south-easterly wind we must make sure the shoots are tucking into the supporting wires to prevent damage.

Wind is a Cape phenomenon. We have two winds that visit us regularly and both can blow with frightening power. The north-wester brings in the cold and rain from the South Atlantic. The south-easter is a dry wind that blows across the coast and into the interior. This wind is known as the Cape Doctor, and this year it is living up to its name, blowing through the vine canopy, keeping it free from disease. As long as the rain stays away, and with the help of the Cape Doctor, there is less chance of a mildew attack. Life can slow down to a ripening pace while the grapes grow plump with juice.

Cherries fill our lives with their bounty for the month of December. I have only one tree, and each year I make a mental note to plant more, but somehow it always slips my mind and I have never managed to do this. But it shall be my New Year's resolution for the coming year. Every December I visit some beautiful cherry orchards found just beyond the pretty Ceres Valley, a short drive from the farm. This is like a pilgrimage that I look forward to each year, and I always invite a friend to come along with me and enjoy the spectacle. This year, I've asked a close friend from our nearest village. She hasn't been before and I am filled with the anticipation of her reaction.

We set out early because I like to arrive at the orchards before any of the other visitors. I confess that I like to absorb the peace and beauty in selfish solitude. As we set off, the mountains are shrouded in light mist. This is a promise of a hot day ahead. The journey passes quickly as we chatter all the way. Soon the car is climbing to a more solitary landscape, approaching rows upon rows of cherry trees. In the orchards the trees are adorned with lipstick-red fruit. The point of this outing is to pick our own fruit, which is then paid for by weight. The trees are so laden they are straining under the weight of their round red fruit. Each year I pick several kilos, to eat over Christmas and make delicious desserts and jam.

The December evenings are light and long. We share suppers on one of the patios, watching the evening turn into night as stars appear from the indigo depths of the sky. Evening swims are pure delight. Fruit bats swoop through the garden with graceful aerobatic skill. As a child I was terrified that this friendly evening visitor might enter a bedroom window. I was ignorant then of their value in the environment. Now I know better and we welcome their nightly visits, particularly during the time when the orchards are flowering. Alongside the bees, bats are one of nature's most important pollinators. In a worldwide context, these fruit-eaters help to spread the forests that are the green lungs of the planet.

The two boys disappear in search of a Christmas tree for the house, which entails driving up to the mountain foothills in search of the perfect specimen. "Not too big!" I cry as they disappear. But I know full well they will return with a monster, strapped to the back of the old Land Rover. While they're gone I make a batch of Christmas mince pies, topped with stars, and a batch of my quick and easy Christmas biscuit dough. Some I cut into Christmas shapes and bake to adorn the tree. The rest will keep in the fridge until I need to bake more.

The tree selected is rather too large, but it has been trimmed down and dragged into the house, leaving a trail of pine needles in its wake. Now it is up to the boys to sort out the decorations and check that the lights are working. I will make a smaller decoration of rosemary to adorn the kitchen window sill. I love candlelight, and often light a candle in the evening. Christmas is a great excuse to go overboard on candles and so I ensure that we have enough of them to fill the assortment of holders scattered around the house. I pin a vine wood wreath on the kitchen door and suddenly the house is filled with the festive spirit of Christmas.

The garden is full of our friends, gathered to share a glass of the best and some food. Paté, cheeses, bread, salads, our home-cured olives and bowls of summer fruits adorn several tables. Under the shade of the huge oaks, a large cool box is topped up with ice and filled with an assortment of bottles. My mother, here from England, gazes adoringly at her two beloved grandchildren as they fill the cooler box with more ice cubes. Three generations of the family sitting around the table sharing stories is a comforting sight.

The long oak table is the focal point of the gathering. It is laden with food and everyone is reluctant to leave its magic; we stay on well into the early hours of the morning. When the children were small I would steal away before midnight and hide. Later, I would ring a small bell and come running out onto the veranda asking everyone if they too had heard the gentle ringing sound. "It must be Father Christmas's sleigh bells!" we'd all say, and the children would shoot off to bed. Disappearing faster than a bullet, they would lie starry-eyed and fearful that he'd seen them up and about, and that their stockings would remain empty.

This naughty trick worked for many years until they clicked that I wasn't around when the bells rang. But I will always remember the excited look on the small children's faces as they galloped from the table. Oh the fun we had with those bells which would never ring until all the desserts and homemade treats had been devoured! These days, as they grow closer to adulthood, those same children laugh at the bell story as they share some cheese and a celebratory glass of champagne. Standing by myself in the silent kitchen I find myself listening to the babble of happy voices outside.

As with every good meal there is always the aftermath. This is dealt with the next morning, with the help of many hungry hands knowing that breakfast can only follow after the mess is cleared. How can we all be thinking about breakfast when only a few hours ago we were feasting on dinner? The day is still and hot as people drift around chatting and nibbling, moving chairs to chase the shade. As evening falls, children play in the pool. Candles are lit as we once again gather around the table. We carry on feasting into the night, reluctant for such a wonderful day to draw to an end.

By the end of Christmas, the farm is silent. The staff are gone for the holidays. The family is sleeping, oblivious to the messy kitchen, which I will deal with later. I walk up to my favourite fruit tree, an old variety of dessert peach, and pluck one from the laden tree. This fruit is so delicate it will bruise under the faintest pressure, making it unsuitable for commercial farming. In our garden orchard it is a gem. As I bite into it, juice squirts in all directions and the taste is heavenly; so perfumed and sweet. I eat one then another.

Their flesh is the perfect temperature in the early morning air, and I carefully pick a full bowl and return to the messy kitchen.

An hour later the rest of the family wakes and stumbles into a now clean kitchen, redolent of the perfume of just-picked peaches, strong coffee and oven-fresh bread. With my chores complete, I decamp to the poolside table with a pile of books and magazines. Sundays are for relaxing and eating simple meals made from leftovers.

Two days later the birds discover the ripe peaches. They tuck in with glee and I am forced to pick more than I need. Suddenly peaches appear at lunch, tossed with rocket, a slosh of olive oil and some goat's cheese, and at supper, made into simple desserts.

There's a colander of slightly unripe peaches sitting on the kitchen table, I cut them in halves and poach them in sweet white wine infused with a vanilla pod, and a sprig of rosemary. Left on the tree to ripen, they would be a feast for birds. Instead, I fill a large jar with fruit and reduce the juices, adding sugar. The syrup is poured over the fruit and the jar is sealed. When I open it someday soon to serve the fruits for a simple dessert, the memories of a happy summer will spread out onto the plate. Damaged fruit is sliced to remove the soiled sections and made into a crumble topped with crushed almonds. I spend a happy afternoon inventing and recording a dozen different recipes for peaches. Being able to "play" in my kitchen, experimenting with new dishes and improving old ones, is one of the luxuries of these lazy December days. I have plenty of volunteers to try them out and the firm favourites are recorded for future use. The garden is my larder and I cherish every day.

I collect some of the early-ripening tomatoes, which I want to prepare for lunch. I scoop out the seeds and stuff them with bread crumbs and herbs, then pop them into a hot oven to bake. The scent that wafts from the oven is mouth-watering. This simple dish was picked up on my travels around Provence in France, where I find the cooks so willing to share their skills. The climate is so similar here that I find many of the Provençal dishes a perfect match for our lifestyle.

It's no fun slaving in a hot kitchen in the middle of the day, so we usually share our holiday feast on Christmas Eve. Depending on what's around, we might have poultry like a duck, a goose or an organic chicken. Of course there's always a special dessert and extravagant Christmas crackers all dressed up in party glitz and filled with delightful little gifts and silly sayings that amuse all the children, young and old.

This Christmas, the feast is succulent roast chicken with whole heads of garlic and rosemary potatoes cooked alongside. I serve everyone a soft head of roasted garlic which they can squash with their fork and spread over their chicken. Garlic, when roasted, has a lovely mild flavour that works so well with the sumptuous taste of our farm-raised organic chicken. We eat one of these delicious birds a week, but it is still a treat for us.

The cherries from my visit to the Ceres orchards have been poached as the finale to this celebratory meal. Cherries are delicious served with macaroons and chocolate ice cream and my boys tuck in and are soon clamouring loudly for more, causing the candles and lanterns to flicker with surprise. We sit around the table sharing stories late into the night. The next day I serve a melon salad with the cold meats for a late Christmas lunch. This is the first melon of the season and I drape it with Parma ham and some mozzarella cheese, delicious and simple.

The period between Christmas and New Year is blissful. Quiet and hot. I'm not a morning person and I struggle to get up early during the year. Perversely now, without all my usual commitments to the family and farm, I wake at six o'clock full of energy. I like to think that it is the luxury of being able to enjoy this beautiful place.

I practically live in a sarong, which is the coolest way to dress for the summer heat that pervades the valley. My most energetic task is picking fruit and vegetables. For fun I try out new recipes on visitors, using up leftovers as I go. Last night's ratatouille is folded into beaten eggs and poured into a pastry case with feta to become a tart. And last night's lamb goes beautifully with a couscous salad with cold roast vegetables.

Lunches seem to last longer, and, served with wine and coffee, are often followed by afternoon siestas under one of the huge old trees. Suppers are eaten much later than usual. No pressure. No hurry. The moon is high in the bejewelled sky before anyone even starts to think of turning in.

A few wine buyers arrive from a frost-covered Europe, mixing business with a sun-filled holiday. They are drawn to the table. There's no nicer way to taste our organic wines than with a good lunch.

Life flows gently into the new year, which is welcomed in on the front patio. One of our first new years on the family farm was the millennium year. We hadn't moved out here permanently yet, but often camped in the half-renovated house at weekends. Then, dark premonitions of computer crashes and chaos covered the front page of newspapers. Midnight slipped quietly by as we sat there on that magnificent evening, sharing some wine and watching the bats swoop by. The mountains that are our backdrop were outlined by a velvet sky and the full flush of stars sparkled like the diamonds of that well-loved children's song. As we retired for the night in the wee, small hours of the new millennium, my mother turned to me and said, "Well the world could be in chaos, but we certainly don't know about it."

In this amazing, peaceful idyll, lulled by nature's majesty and lacking a radio or a TV, we were oblivious to the outside world.

Roast Chicken with Whole Heads of Garlic

2	large organic or free-range chickens
	large knob of butter
6	large roasting potatoes
6	heads of garlic
	drizzle of olive oil
2	sprigs of rosemary
	sea salt

Serves 6

Rub the chicken with softened butter, and place in a roasting tray. Place in a preheated oven at 190°C or gas mark 5. Peel and cut the potatoes into cubes, place in a pan of boiling salted water, simmer for 5 minutes to parboil them, then drain. Remove the chicken after 30 minutes and add the garlic heads. Drizzle a little olive oil over each head, then add the potatoes and 1 sprig of rosemary and return to the oven for 1 hour. Take the tray out after 30 minutes to turn the potatoes and garlic over. The potatoes will roast in the melted butter and juices from the chicken. Remove from the oven, and the juices should run clear from the chicken. Remove the chickens from the pan and let them rest for 5 minutes. Carve and arrange the chicken on a platter and remove the potatoes and garlic from the oven. The potatoes should be pale golden brown and the garlic soft. Place the potatoes and garlic around the chicken on the platter, spoon some of the juices from the pan over the chicken, then sprinkle with more sea salt and roughly torn rosemary from the remaining rosemary sprig.

Serve each person a soft head of roasted garlic which they can squash with their fork and spread over their chicken. Bear in mind that garlic when roasted has a much milder flavour.

Melon Salad

1	sweet melon
100g	rocket leaves
150g	mozzarella cheese
12	slices of air-dried pancetta. Parma ham will also work.
	drizzle of olive oil
	drizzle of balsamic vinegar.

Serves 6   Cut the melon in half. Peel it and remove the seeds. Spread the rocket leaves onto a platter, then cut the melon into thin slices and place them on top of the leaves. Cut the mozzarella into rough slices, then add to the platter. Lastly, lay over the pancetta slices, drizzle over the oil and vinegar, and serve.

Poached Cherries

50g	unrefined white sugar
125ml	water
750g	cherries, stoned
	dash of brandy (optional)

Serves 6   Place the sugar and water into a pan and bring to the boil. Turn down to simmer and add the cherries. Simmer for 5 minutes. If adding brandy, do so at the end of the cooking time. Cool, then place in the fridge. Serve cold.

Tip: Serve with cream or ice cream and macaroons.

Macaroons

	makes about 40 macaroons
4	egg whites at room temperature (not straight from the fridge)
125g	almonds, finely ground
200g	icing sugar
1 tbsp	plain flour
	dash of vanilla essence
	sheets of rice paper

Beat the egg whites in a bowl, until they are stiff enough to hold their shape. Fold in the almonds, sugar, flour and vanilla with a metal spoon. Fold gently so the air you have just beaten into the eggs isn't lost. Place the mixture into a piping bag with a wide nozzle. (If you don't have a piping bag, spoon the mixture onto the rice paper with a teaspoon.) Pipe balls of the mixture onto rice paper placed on a baking tray. You may have to cook them in 2 batches. Place the biscuits in a preheated oven at 160°C or gas mark 3 for 5–8 minutes. The macaroons should be firm to the touch. Leave them on the tray for 5 minutes before you transfer them to a cooling tray. You can leave them single, or while still warm attach 2 by gently pressing the flat sides together. The rice paper will stick firmly and is edible. They will keep in an airtight container for 2 to 3 days.

Lemon & Almond Christmas Cake

	makes 1 cake
200g	unsalted butter
200g	unrefined white sugar
2	eggs
1	large lemon, zest and juice
200g	ground almonds
200g	plain flour
1 tsp	baking powder

Beat the butter and sugar together until creamy, add the eggs 1 at a time, then the lemon zest and juice. Stir in the ground almonds and sift in the flour and basking powder. Place in a cake tin lined with greased baking paper (keep butter wrappers folded up in the fridge to grease baking paper when needed) and bake at 180°C or gas mark 4 for 45–50 minutes. Test by inserting a skewer – if it is clean when removed the cake is cooked through. Leave it to cool for 15 minutes, remove from the tin and leave to cool totally on a rack. This cake is blissfully plain – for a festive touch decorate with glacé or dried fruits and nuts.

Christmas Biscuits

Makes about 50 biscuits depending on the cutters

125 g	butter
50g	unrefined white sugar
50g	organic brown sugar
2	eggs
	dash of vanilla essence
50g	golden syrup
350g	plain flour

Place the butter and sugar into a mixing bowl and beat together until creamy. Add the beaten eggs, vanilla and golden syrup, and lastly fold in the flour. Place the dough into the fridge to rest for 20 minutes before rolling and cutting out your shapes. Roll out the dough to roughly 1cm thick and cut out your chosen shape. Remember, if you want to hang the biscuits on a ribbon as a decoration, cut the hole for the ribbon before baking – cut it slightly larger than you require and not too close to the edge of the dough. If using for decorations, these biscuits can be iced, but also taste great on their own.

This is a good dough recipe to use if you want to make biscuit decorations as it is not too crumbly and has a nice crunch and firm texture. I make a large batch by doubling the specified amounts and keep it in the fridge for up to 2 weeks.

List of Recipes